A Guide to
Southern Temperate Seagrasses

Michelle Waycott, Kathryn McMahon, Paul Lavery

CSIRO

PUBLISHING

National Library of Australia Cataloguing-in-Publication entry

Waycott, Michelle, author.

A guide to southern temperate seagrasses/Michelle Waycott, Kathryn McMahon and Paul Lavery.

9781486300150 (paperback)
9781486300167 (epdf)
9781486300174 (epub)

Includes bibliographical references.

Seagrasses – Southern Hemisphere.
Seagrasses – Southern Hemisphere – Identification.

McMahon, Kathryn, author.
Lavery, Paul Stephen, author.

584.701814

Published by
CSIRO PUBLISHING
36 Gardiner Road, Clayton VIC 3168
Private Bag 10, Clayton South VIC 3169
Australia

Telephone: [+613] 9545 8555
Local call: 1300 788 000 (Australia only)
Fax: +61 3 9662 7555
Email: csiropublishing@csiro.au
Web site: www.publishing.csiro.au

Printed by Ingram Lightning Source

CSIRO PUBLISHING publishes and distributes scientific, technical and health science books and journals from Australia to a worldwide audience and conducts these activities autonomously from the research activities of the Commonwealth Scientific and Industrial Research Organisation (CSIRO). The views expressed in this publication are those of the author(s) and do not necessarily represent those of, and should not be attributed to, the publisher or CSIRO. The copyright owner shall not be liable for technical or other errors or omissions contained herein. The reader/user accepts all risks and responsibility for losses, damages, costs and other consequences resulting directly or indirectly from using this information.

Facing page left: *Posidonia australis*.
Facing page right: *Posidonia ostenfeldii* 'complex'.

Feb26_RP_ILS

I am delighted to provide the foreword to this important contribution that provides new insights through summarising the advances in our knowledge of the distribution of southern seagrasses. It is a beautiful book.

Since Professor Cornelis den Hartog's 1970 The sea-grasses of the world, many more researchers have explored deeper waters, both on SCUBA and by remote sensing, than were envisaged when den Hartog first relied on intertidal collections of drift material to infer distributions. Southern Australia's seagrasses have also benefitted from major taxonomic work by the late Professor Bryan Womersley and Enid Robertson in Adelaide, where Professor Michelle Waycott is now based, and from the detailed anatomical considerations by Emeritus Professor John Kuo, at The University of Western Australia in Perth. Emeritus Professor Arthur McComb, also at UWA and Murdoch University, played a vital role in training the present generations of seagrass biologists and ecologists in southern Australia, including the authors of this book.

The significance of seagrasses as real flowering plants has developed as our ability to study them underwater has also developed. Advances in molecular biology have also led to a clearer evolutionary picture of how the different plants are related.

Increasing awareness of seagrasses as important habitats is vital, as seagrasses are on the front line of rising sea levels and increased storminess associated with climate change. We are losing these 'coastal canaries' at an alarming rate. Maintenance of healthy seagrass beds into the future will rely on us developing that awareness, and this book will certainly help.

Emeritus Winthrop Professor Diana I Walker
11th June 2013

Many requests to extend the seagrass guides series to southern temperate species were made following the success of A guide to the tropical seagrasses of the Indo-West Pacific and encouraged this endeavour. On embarking on this task, the three of us, Kathryn, Paul and I, concentrated on the seagrasses of southern Australia. However, we quickly realised that adding the seagrasses of New Zealand, South America and southern Africa wouldn't expand things dramatically, so we included them. We did not envisage some of the hurdles we would need to overcome: the birth of baby Pearl, significant illness, several job changes including moving interstate, significant taxonomic debate and changes in our graphic design and technical support. Despite this, we are proud of the significant advance this guide provides to anyone wanting to look at and understand more about the seagrasses from the southern temperate coastal marine ecosystems of the world.

Taxonomy is a living, dynamic field of study, not without its points of conflict, particularly for seagrasses. There are several places in this guide where I have taken an alternate view to other taxonomic works with respect to current species concepts. I take full responsibility for these decisions and try to explain simply the rationale and encourage users of this guide to a greater level of awareness.

I wish to give particular thanks to several people: first and foremost Ainsley Calladine, who helped with many of the graphics, was a source of technical advice, publishing know-how and provided encouraging appreciation of seagrasses to one and all; to Tirza Abb, our graphic design support, thanks for your patience and skill; Diana Kleine, for your initial graphic design and layout input and encouraging the ongoing production of the series; Bill Dennison, as always your enthusiasm was important for putting good science into widely accessible forms and ideas; finally to Don Les, for your intellectual support and critical ideas and finely honed humour.

Michelle Waycott, Adelaide, SA, Australia, May 2013

Opposite: Orange ascidian growing on *Posidonia*.

Contents

Opposite: Starfish in *Posidonia*.

Left: Leaf clusters and stems of *Amphibolis antarctica* seagrass with algae attached. Top right: A box fish in *Halophila* seagrass. Bottom right: Long strap-like leaves of *Posidonia* seagrass growing with *Caulerpa*, a green alga.

What is a seagrass?

Seagrasses, also known as marine flowering plants or angiosperms, occupy the coastal marine habitats of all continents except Antarctica. They are common in shallow waters, typically down to about 20 m in depth, where there is sufficient light to photosynthesise and a suitable surface to grow on. In some places, seagrasses grow where they are exposed to the air during low tide but elsewhere can grow down to 70 m water depth.

Like all flowering plants, seagrasses produce flowers and seeds, and are more closely related to giant redwood trees than the marine algae (seaweeds), with which they coexist in marine ecosystems. Algae originated before the dinosaurs and while some may superficially resemble seagrasses, there are several differences. Unlike algae, seagrasses have veins or vascular tissue in the leaves, true roots and flowers. While many have simple strap-like leaves, similar to grasses, others have stems with complex arrangements of leaves on them.

People who live near the coast may be aware of marine organisms associated with seagrass meadows, such as rock lobster or seahorses, but may be unaware that seagrass meadows are nearby. The tell-tale signs of a seagrass meadow are the leaves and stems that wash up on the beach (beach wrack). In the water, seagrass meadows often have long, waving leaves which can completely cover the sediment with a diverse array of fish and small crustaceans in and around them.

♀

Halophila australis Doty & Stone
Farm Beach, Coffin Bay, S.Au
5m deep
3.XI.1981
coll. S.A. Shepherd
Det. E.L. Robertson

Herbarium specimens are important records for the field of taxonomy. This specimen is *Halophila australis*.

Taxonomy

Taxonomy is the science of finding order among groups of organisms and, from this, developing a classification. Historically, taxonomy was based on morphological and reproductive features. The 20th Century has seen improvements to the data available for taxonomists to use, from DNA barcoding and molecular phylogenetic analyses. For seagrasses, taxonomy is sometimes problematic as they have highly reduced morphologies, which may also be very variable.

Early classifications of seagrasses placed them with various groups of algae. In the mid-1800s they were finally recognised as 'flowering plants'. The most influential taxonomic work on seagrasses has been the treatise entitled *The sea-grasses of the world* by Cornelis den Hartog (1970). Since then, and despite new data generated from modern analyses, the number and composition of genera remains largely unchanged. Taxonomy updates, including genera and species names and all the previous names used for each species, are listed on the International Plant Name Index website (www.ipni.org). Throughout this guide we use the family and genera names currently approved by the Angiosperm Phylogeny Group (APG III).

Defining species in some seagrass genera remains difficult. Often, the species within these genera lack distinguishing features and while some plant attributes do vary, for example leaf width, this variability is overlapping among species, making discrimination very difficult. Recent genetic analyses have found some of these taxa are not distinctive e.g. several previous Australian and New Zealand species of *Zostera* are now recognised as one, *Z. muelleri*. Ongoing taxonomic ambiguity exists for the *Posidonia ostenfeldii* 'complex', and the *Halophila ovalis* 'complex' where characters overlap and molecular analyses do not support current taxonomy. Also, species of *Ruppia* and *Lepilaena* are commonly found in brackish waters, estuaries and saline lakes, but it remains uncertain which species are principally marine and we unreservedly treat the marine members of these genera as 'seagrasses'.

In this guide, we have largely followed the classifications of Cronquist (1981) and den Hartog (1970). However, in several places we have pointed out where more recent work has created uncertainty regarding the species boundaries, in particular for the genera *Halophila*, *Posidonia* and *Zostera*. More details on these uncertainties are presented in the family and species pages. The guide does not seek to resolve these taxonomic uncertainties but to bring them to the attention of readers. In the majority of cases, ongoing research at a global scale is underway to resolve these issues.

Habitats

Four general types of seagrass habitat are readily identified in southern temperate environments and are similar to other parts of the world: 'Estuarine', 'Coastal', 'Reef' and 'Deep'. Estuarine, coastal and reef habitats can also be categorised based on how they are influenced by tides (either intertidal or subtidal), while coastal and reef habitats can be categorised by their exposure to wave energy (sheltered or exposed). Many deepwater habitats are exposed but the wave energy is reduced before reaching the seagrasses growing on the bottom. Different combinations of conditions exist in each of these ecosystems and, because each seagrass species has specific tolerances for different environmental conditions, different species of seagrass occur in each location.

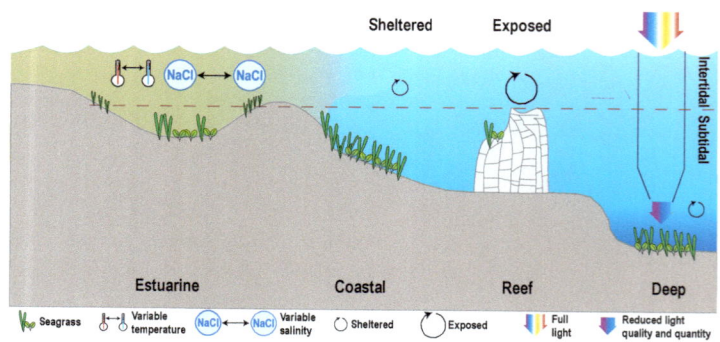

• Intertidal

Habitats that are exposed to air during at least some low tides. Species tolerant of desiccation and a wide range of temperatures survive in these habitats.

• Subtidal

Habitats that are never exposed to air, the deeper the water the less light reaches the plant.

• Sheltered
Habitats protected from ocean swells and strong prevailing winds, with low sediment movement.

• Exposed
Habitats without protection from ocean swells and strong winds, with high sediment movement.

• Estuarine
Highly dynamic environment, variable river flow and ocean openings, fluctuations in salinity, temperature, light and nutrients, often high sediment deposition.

• Coastal
Soft, often sandy sediment with annual variations in water temperature, relatively consistent salinity and often high input of nutrients, variable water clarity.

• Reef
Limestone, granite or basalt reef with thin soft sediment veneer, annual variations in water temperature, relatively consistent salinity and high wave energy.

•Deep
Soft sediments or reef in greater than 20 m of water depth, with low light and low wave energy.

Top left: Algal and ascidian epiphytes. Top right: Sea lion over seagrass. Bottom left: Seagrass fibre ball. Bottom right: Gastropod grazing.

Ecology

Temperate seagrasses introduce a diverse set of complex habitats to coastal waters. They are an important foundation species, strongly influencing the structure of coastal marine communities. Seagrasses, and the epiphytic algae which grow on their leaves and stems, provide habitat and/or food for a wide range of animals. The magnitude of their influence is driven by the size and meadow-structure of the seagrass. Larger, meadow-forming species provide critical habitat for a wide variety of fauna, including some fish which are important for fisheries, and invertebrates such as prawns, crabs and starfish. But even the smallest seagrasses, which form only sparse meadows, play a similar role.

The diverse array of organisms inhabiting seagrass ecosystems perform critical ecological functions. These seagrass communities filter the water, removing and recycling nutrients, and also stabilise sediment on the seabed. The epiphytes growing on seagrasses are important food for many grazers, though southern temperate seagrasses themselves are not usually a direct food source for grazers. This contrasts with tropical seagrasses, which are regularly eaten by fish, urchins, dugongs and turtles. However, in some southern temperate areas, swans depend on seagrasses for food.

Seagrasses play a crucial role in connecting different marine habitats. The meadows are a fertile hunting ground for predators who live in nearby reefs. Algae that are detached from reefs during storms drift into seagrass meadows where they release nutrients that support seagrass and algal growth. In turn, seagrasses and their epiphytes break off and drift to other habitats, such as beaches, where they are an important source of food and habitat for a variety of animals. The accumulations of detached leaves can also help to stabilise beaches, preventing erosion during storm events.

Top left: Blue manna crab. Top right: Polycheate worms. Bottom left: Nudibranch. Bottom right: School of trumpeter fish.

Fauna

Even the most cursory swim over a seagrass meadow reveals a wonderful diversity of fauna in, on or under the seagrass canopy and its sediments. The fauna is typically dominated by small crustaceans and molluscs, but seagrass meadows also support larger fauna such as fish, crabs, clams and swans.

Large seagrass meadows generally have the most abundant and diverse fauna. For example, an *Amphibolis* meadow can support more than 100 species of invertebrates and 70 species of fish. Some of the associated fauna have important commercial or recreational value; the western rock lobster fishery in south-western Australia is valued at hundreds of millions of dollars each year. Even sparse seagrass can support dozens of species of fish and invertebrates.

The Black Swan is associated with temperate seagrass meadows and commonly feeds on *Ruppia* and *Halophila* in estuaries of southern Australia. The loss of *Ruppia* in the Coorong, South Australia, has been linked to the decline of waterbirds in this RAMSAR site. In Shark Bay, close to the northern limit of the Australian temperate seagrass species, the world's largest dugong population feed on several seagrass species, including *Amphibolis*. In southern Australia, the leafy sea dragon is well camouflaged in the seagrass it inhabits as are pipefish, whose elongated bodies mimic seagrass leaves, helping them to avoid predators but confining them to seagrass meadows. At least 74 species of concern on the IUCN Red List are associated with seagrass habitats and the survival of seven species of threatened shark has been linked to the health of seagrass meadows.

Top left: Seagrass wrack on beach. Top right: Research. Middle left: Seagrass beach wrack can be dangerous. Bottom left: Leaves bleached by high temperatures. Bottom right: Underwater experiments.

Southern temperate seagrasses have been used by humans for hundreds, if not thousands, of years. Aboriginal Australians used fibres of *Posidonia* to weave cloaks and baskets. The same fibres were commercially harvested in South Australia in the early 1900s for insulation, stuffing material and to mix with low-grade wool in coarse textiles.

Historically, indigenous peoples harvested seafood from seagrass meadows and today people still seek out meadows as productive fishing grounds. But the greatest benefit humankind derives from these ecosystems are indirect, resulting from the ecosystem services they perform. These services include their support of coastal fisheries, coastline stabilisation and the capture of carbon from the atmosphere. Economists have valued the benefits that humans derive from seagrass meadows at between $15,000–$20,000 per hectare per year.

Despite the benefits we derive from seagrasses, human activity has caused serious damage to seagrass habitat. Direct loss has occurred through dredging, construction and boat moorings. Indirect losses have resulted from nutrient or sediment pollution of waterbodies. Some of the world's largest losses have occurred in the southern temperate bioregion (e.g. Cockburn Sound, WA, and Western Port, Victoria). Major losses have also occurred in South Africa, mostly due to sediment run-off into estuaries. Little is known about seagrass loss in South America.

The extensive loss of these valuable ecosystems has triggered research, monitoring and management programs to improve the understanding of seagrasses, track changes in habitat and manage human impact. Globally, networks of seagrass monitoring have been established, such as Seagrass-Watch and SeagrassNet, and increasingly seagrasses are used as early warning indicators of environmental degradation—'canaries of the seas'. Significant research has been undertaken into seagrass rehabilitation and transplantation techniques, particularly for *Posidonia* spp. The results are varied, but at this stage large-scale transplantation is neither technically nor financially feasible.

Top (left to right): *Posidonia australis* flowering shoot, inflorescence and fruit.
Bottom (left to right): *Zostera tasmanica* flowering shoot, inflorescence showing male flowers, female flowers and seeds and a seedling.

Reproduction

Seagrasses grow and reproduce by two methods, similar to some grasses on the land. The first method is clonal growth, where new leaves, underground rhizomes and roots are produced from the existing plant. Larger seagrasses, like *Posidonia,* are relatively slow growing, with one shoot producing around two leaves per year, whereas smaller seagrasses with different growth forms, such as *Halophila*, can produce up to 180 leaves per year from a single shoot.

The second method is sexual reproduction; seagrass plants flower, pollinate and set seed, and the seeds germinate, leading to the establishment of new plants. Most seagrasses have male and female flowers on separate plants, though some have male and female flowers on the same plant (e.g. *Zostera*). *Posidonia* is unique in the marine world with its bisexual flowers that have both male and female parts, a characteristic of most land plants. Generally, seagrass flowers are inconspicuous, as they are small and are held within the canopy. However, in some species the flowers are more noticeable, such as in *Posidonia* with its specialised shoots that bear flowers or *Ruppia*, which sends flowers to the water surface on a spiralled stalk, with pollination occurring just above the water.

The type of seed varies, depending on the genera. For example, some genera, such as *Zostera,* form a seed with a hard outer covering, whereas *Posidonia* seeds have no hard covering, and start to grow immediately after release from the parent plant. A less common form of seed development, termed 'vivipary', occurs in *Amphibolis* and *Thalassodendron*; the new individual develops on the adult plant and is released as a small seedling.

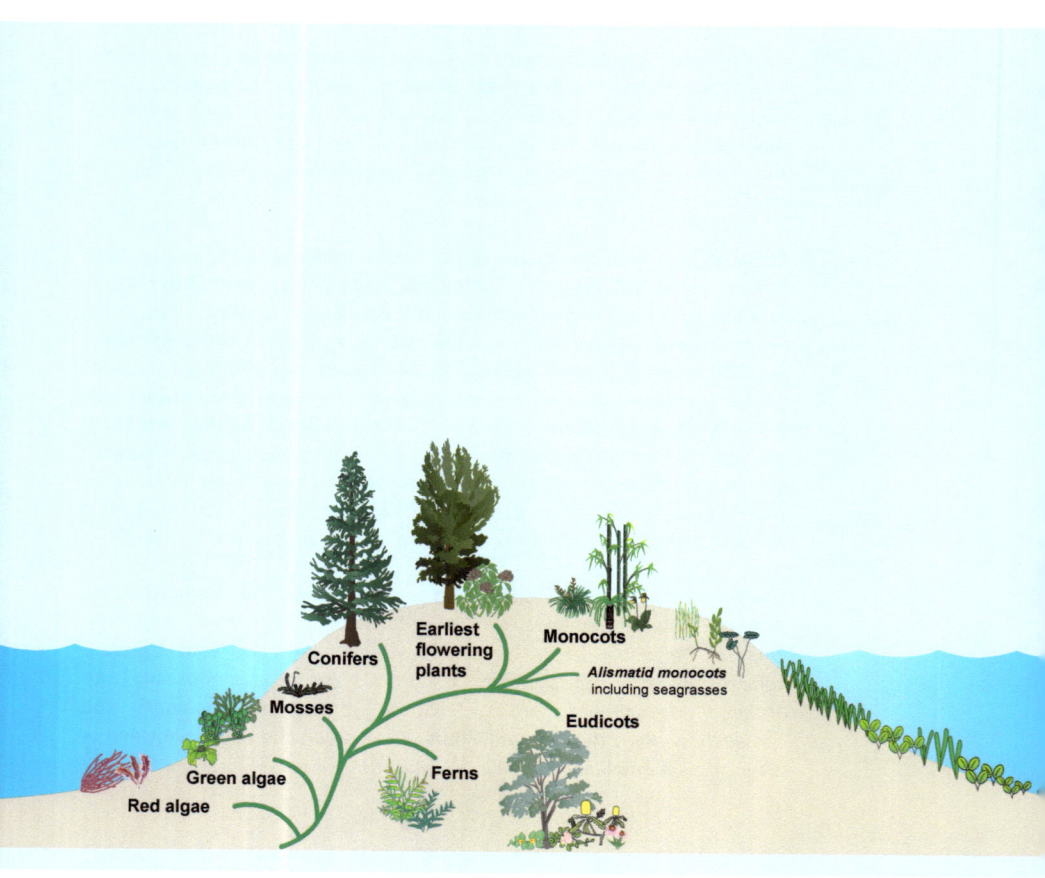

Depiction of our current understanding of seagrass evolution. The ancestors to seagrasses moved onto land from the sea around 400 million years ago. They subsequently diversified into a number of groups. Mosses, ferns then conifers evolved via a series of steps over the next 300 million years. The flowering plants, including the monocots, originated more than 120 million years ago. Some of the monocots evolved and returned to the sea in the form of seagrasses around 100 million years ago.

Evolution

The oceans of the world are full of 'plants' that live their whole life in or under the sea. These marine plants fall into four groups: phytoplankton (including diatoms), algae, mangroves and seagrasses. Today we recognise more than 60,000 species of marine plants, of which 50,000 are phytoplankton, around 10,000 are algae, 70 are mangroves and around 70 are known as seagrasses.

Millions of years ago flowering plants, or angiosperms, evolved from 'plants' that were living in the ocean and became adapted to life on the land. There are more than 250,000 modern angiosperms. Seagrasses are angiosperms, which means they have land-based ancestors and have re-adapted to the marine environment. Today, four different groups of seagrasses have adapted to life in the ocean, some with freshwater aquatic relatives (family Hydrocharitaceae) and some with saltmarsh and freshwater relatives (families Zosteraceae, Potamogetonaceae and the Cymodoceaceae 'complex').

A significant proportion (more than 40%) of the global diversity of seagrasses exists in the southern temperate oceans. Among them are species from ten of the 13 genera found throughout the world and, among these genera, three are typically found in tropical warmer waters but make their way into cooler temperate waters in a few regions. The seagrass flora of south-western Australia is particularly diverse, reflecting the convergence of its regional oceanography with an evolutionary ecology which is unique in a global context.

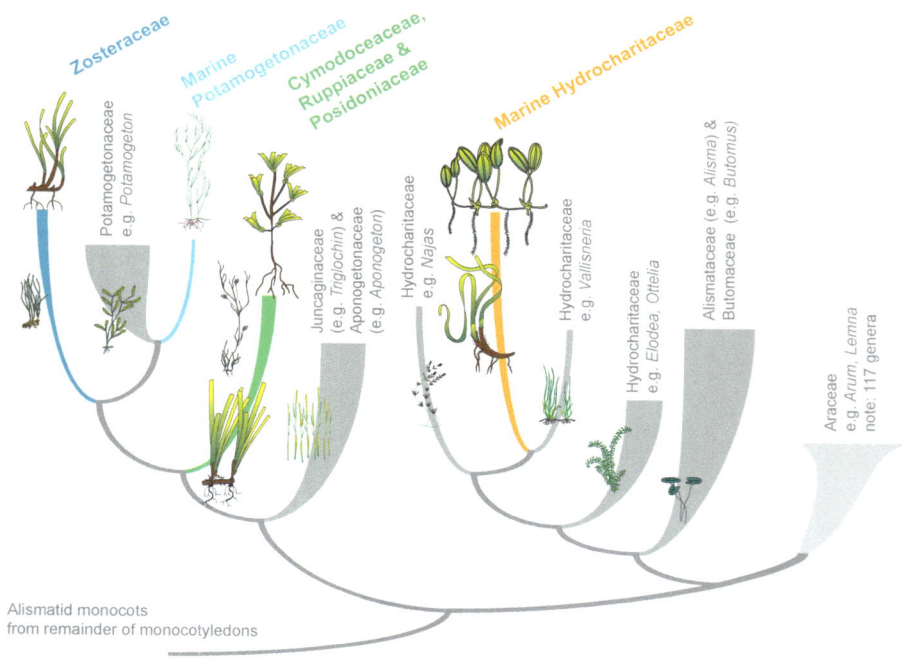

Zosteraceae

Marine Potamogetonaceae

Cymodoceaceae, Ruppiaceae & Posidoniaceae

Marine Hydrocharitaceae

Potamogetonaceae e.g. *Potamogeton*

Juncaginaceae (e.g. *Triglochin*) & Aponogetonaceae (e.g. *Aponogeton*)

Hydrocharitaceae e.g. *Najas*

Hydrocharitaceae e.g. *Vallisneria*

Hydrocharitaceae e.g. *Elodea, Ottelia*

Alismataceae (e.g. *Alisma*) & Butomaceae (e.g. *Butomus*)

Araceae e.g. *Arum, Lemna* note: 117 genera

Alismatid monocots from remainder of monocotyledons

A stylised evolutionary tree for the aquatic plant order, Alismatales (alismatid monocots) based on our current understanding of relationships among the families of this order following the Angiosperm Phylogeny Group III classification. The four major lineages of seagrass plant families are highlighted in colour. The width of branches reflects the number of currently recognised species in each group except for the Araceae, which has more than 4000 species, eight times the number of species in the remainder of the alismatid monocots.

The origins of seagrasses occurred around 100 million years ago, a time when the Earth looked very different; dinosaurs existed and there were very few flowering plants at all. At that time, among the earliest groups of flowering plants was the lineage that became what is today known as the Alismatales, an order of flowering plants containing more than 2,500 species. All the seagrasses occur within this order, the majority of other species in this order are aquatic plants or occur in aquatic ecosystems.

Well-known non-seagrass members of the Order Alismatales include the aroids and duck weeds (Araceae), arrowgrass (Juncaginaceae) and pondweed (Potamogetonaceae). Many of them are globally distributed water weeds, including *Sagittaria* (water plantain) and *Pistia* (water lettuce). The members of this order have repeatedly evolved a range of adaptations to environmental conditions, including submerged pollination (hydrophily), marine salinities (osmoregulation and anatomy) and fluctuations in environmental conditions (clonality). The suite of adaptations required for flowering plants to survive fully submerged in the marine environment place strong constraints on evolutionary diversification into this habitat.

Molecular analysis of relationships among members of the Alismatales by Don Les and collaborators indicates that there have been four independent origins of seagrasses that have the requisite adaptations enabling their survival as submerged marine plants (see figure on opposite page). The four lineages are: 1. marine Hydrocharitaceae; 2. a group of three families, the Cymodoceaceae, Ruppiaceae and Posidoniaceae; 3. the Zosteraceae; and 4. the Potamogetonaceae. These lineages are interspersed between groups that are freshwater aquatics, marsh plants, estuarine species and terrestrial plants.

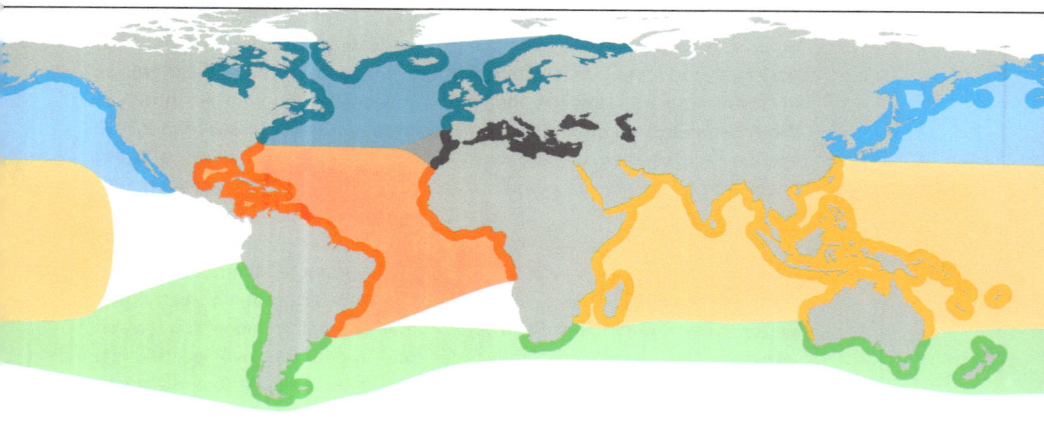

The six seagrass bioregions defined by the suite of species that live there.

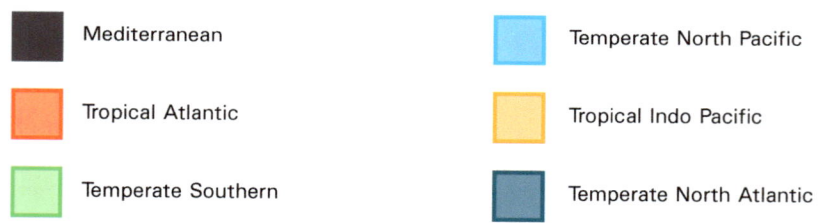

Mediterranean

Tropical Atlantic

Temperate Southern

Temperate North Pacific

Tropical Indo Pacific

Temperate North Atlantic

Bioregions

Different seagrass species group together in different regions of the world. Broadly, seagrass species are either temperate (cooler water) or tropical (warmer water). Some genera are almost exclusively temperate, such as *Posidonia* and *Amphibolis*. Tropical genera include *Enhalus*, *Thalassia* and *Halodule*. There are also genera whose species occupy both temperate and tropical marine waters, such as *Halophila* (the paddleweeds), *Cymodocea* (found in the Mediterranean and the Indo-West Pacific Oceans) and the estuarine species of *Ruppia* (tasselweed, found all around the world).

Seagrasses have recently been grouped into six 'bioregions' where similar sets of species occur. The Southern Temperate Oceans bioregion includes the three southern temperate continents, Australia, South America and Africa, as well as New Zealand. In Australia, temperate seagrasses occupy the cooler, southern waters, from Shark Bay on the west coast (though some species of *Posidonia* and *Amphibolis* occur as far north as Ningaloo Reef) to northern NSW and including Tasmania. The west coast contains a large region where temperate and tropical species overlap due to the Leeuwin Current bringing warm tropical waters into temperate regions. In South America, there are few records of seagrasses; estuarine seagrasses occur on the east coast and a small region of the Chilean coast. In southern Africa, there are no definitive records of seagrasses on the temperate west coast, however, a few species occur in sheltered bays and estuaries of South Africa and more species in the subtropical waters of the east coast, where tropical and temperate species co-occur. Three species of temperate seagrasses have been recorded in New Zealand, all widely distributed across both Islands but generally restricted to estuaries, sheltered harbours and inland waters.

Top: Major currents of the Southern Hemisphere, including the three eastern boundary currents.
Bottom: Floating fruit of *Posidonia* sp. representing an important dispersal mechanism for members of this genus.

Currents

There is a startling contrast in seagrass species diversity in the temperate region of the three southern continents; Africa (10), Australia (28) and South America (3). Ocean currents along these shores play a part in limiting the distribution of seagrasses as they influence ocean temperature, affecting seagrass survival. Currents also transport seagrass plants, fruits and seeds.

The most obvious limits to continental seagrass distributions occur on the west coast of South America and Africa, where there are very few species. The Humboldt Current in South America and the Benguela Current in South Africa move cool water northwards along the west coasts. There are also cold-water upwellings in these regions, which bring cold, nutrient-rich water to the surface. Consequently, only temperate species are found on these coasts as it is too cold for tropical species. In contrast to the western coast, the warm-water Agulhas Current flows southward down the east coast of Africa, allowing a high diversity of both tropical and temperate species to occur there.

Australia differs from Africa and South America as warm currents move south down both the east (East Australian Current) and west (Leeuwin Current) coasts, as well as a transient north-moving, cold counter-current on the west coast. This results in temperate species being found further north on the west coast than on the east coast and a large overlap of temperate and tropical species; it is one of the areas of highest seagrass biodiversity in the world.

Top left: *Posidonia coriacea* growing in high-energy habitats of southern Australia. Top right: Seahorse over *Amphibolis* seagrass in Jervoise Bay. Bottom: *Amphibolis* is endemic to southern Australia.

A unique flora

The seagrass flora of Australian marine waters has a particularly diverse and magnificent range of species; 36 species out of 72 species worldwide, and all but one of the genera. The presence of so many species is partly due to Australia spanning both tropical and temperate bioregions but also because of the abundance of endemic species. These endemics include the Australian *Posidonia* species and both *Amphibolis* species. This level of endemism is unusual as most seagrass species have broad geographic ranges; *Zostera marina*, arguably the best known seagrass around the world, grows across virtually the whole of the Northern Hemisphere in temperate waters.

Amphibolis is the only genus endemic to a single continent, and represents a remarkable adaptation to the cooler water, higher energy (i.e. waves) environments of southern Australia. The closest relatives of *Amphibolis* are predominantly tropical: *Thalassodendron*, *Cymodocea* and *Syringodium*. The Australian *Posidonia* species are all endemic and include a group, the *Posidonia ostenfeldii* 'complex', adapted to survive in southern ocean waters exposed to highly mobile sands and waves.

This diversity has likely evolved through the geographic isolation of the Australian continent. Australia was part of the supercontinent Gondwana during the time of the dinosaurs, but 60 million years ago continental drift moved it away from the other 'southern continents'. Like kangaroos and koalas on the land, new species of seagrasses are likely to have evolved during this long period of isolation. In addition, new species may have evolved through adaptation to a particular environment. Both *Amphibolis* species and members of the *Posidonia ostenfeldii* 'complex' show adaptations to survive in high-energy environments.

Using this guide

The key: Follow the steps

STEP 1

Choose the leaf form, on one of the two pages

STEP 2

Within the leaf form, make choices following the green arrows, working down the page, until you reach a species name.

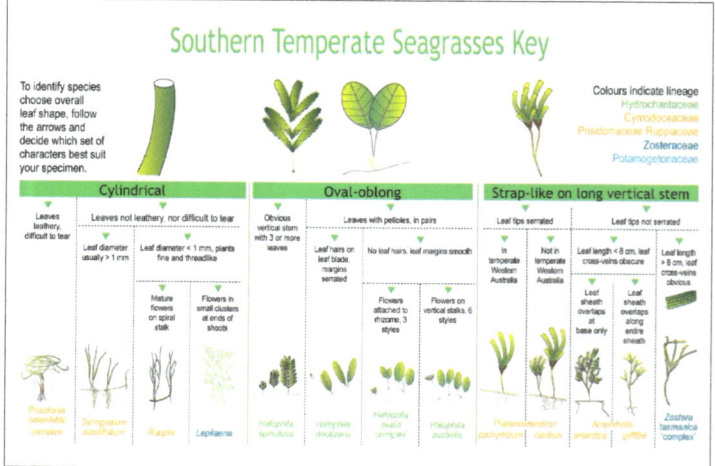

Southern Temperate Seagrasses Key

STEP 3

Go to the next page [species list] to confirm the key characteristics and directions to species descriptions, which are ordered by lineage, family, genus and species. Some species are grouped into a 'complex' as their taxonomy is currently under investigation.

Description: Colours indicate seagrass lineage

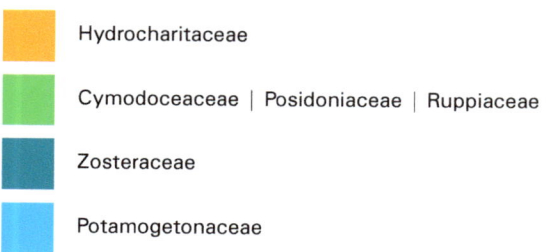

Hydrocharitaceae

Cymodoceaceae | Posidoniaceae | Ruppiaceae

Zosteraceae

Potamogetonaceae

Species icon shows the general form.

The species page

Includes information on the ecology, taxonomy, reproduction and distribution, as well as key identifying features.

▼ Banner *Recognised botanical name*   ▼ Family name

| *Posidonia australis* | Hooker f. | Posidoniaceae |  *Posidonia australis* |

▲ botanical authority

▶ General Form: Showing the general form of the plant in diagram format

▶ Seed: This section shows the main reproductive propagule (fruit, seed, seedling) you are likely to see

Seed

Habitats

▶ Habitats: Coloured habitat icons indicate habitats the species is usually found in

▼ Reproduction: The bars show the months over which flowering and fruiting have been observed in the temperate Southern Hemisphere. Due to the wide distribution of some species and the limited data available, exact timings may vary. The absence of colour indicates flowering has not been observed. For some species the timing of flowering is unknown and the entire bar will be white

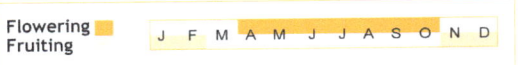

| Flowering ▮ | J F M A M J J A S O N D |
| Fruiting | |

▼ Vulnerability Status: VU Vulnerable, LC Least concern, NT Not threatened, DD Data deficient, as assessed by the IUCN www.iucnredlist.org

NT

▼ Distinctive feature: Helps to distinguish from similar species

Distinctive feature

▶ Leaves < 7 mm wide, tips frayed; old sheath hairy

▼ Distribution: does not imply a species is found in all locations within the shading, but that if conditions are suitable, it might occur there

NT

Key characteristics

▶ Leaves 4–7 mm wide
▶ Intact leaf tips oblique
▶ Rhizome usually laterally compressed

◀ Key characteristics: Helps to distinguish from all other species

Southern Temperate Seagrasses Key

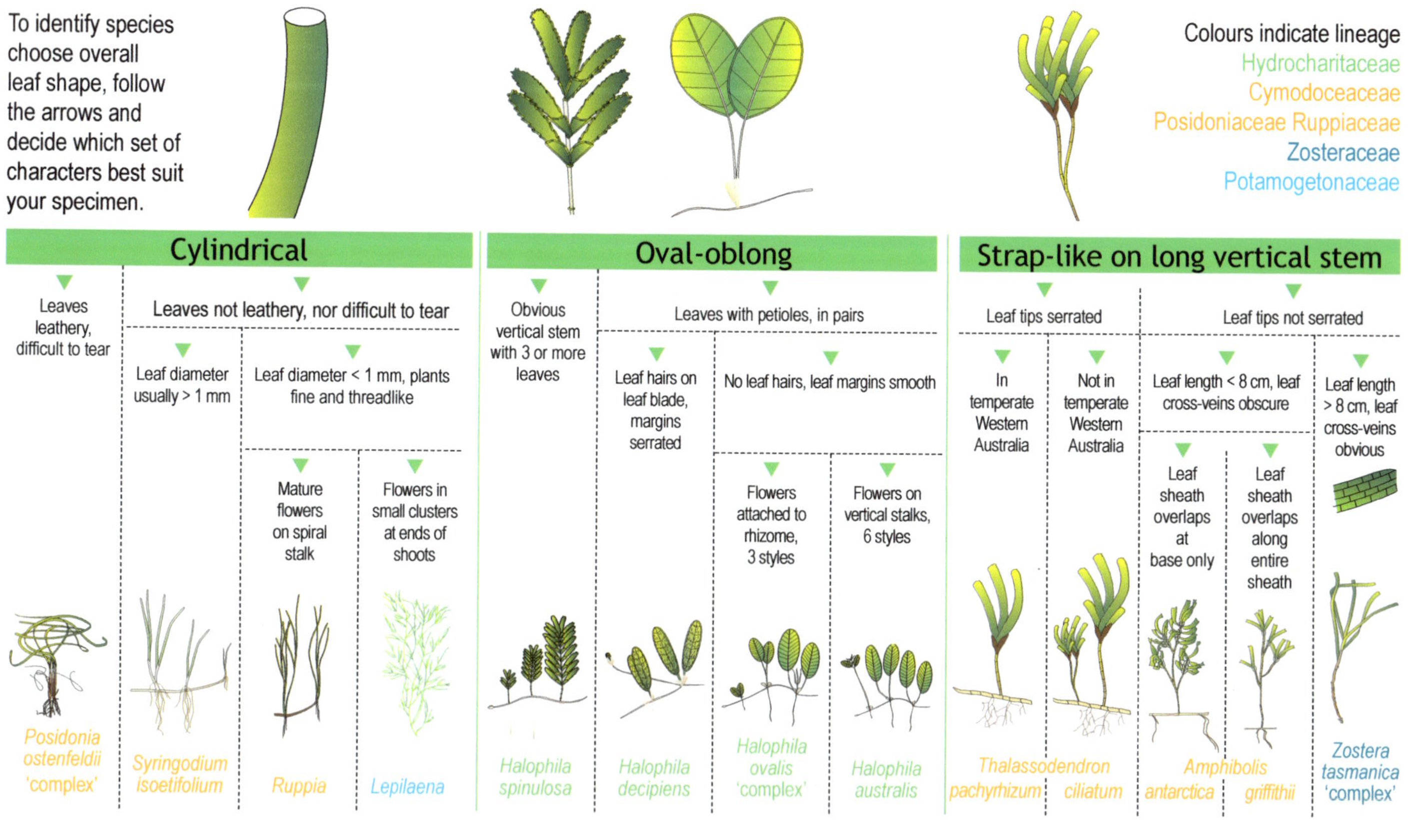

To identify species choose overall leaf shape, follow the arrows and decide which set of characters best suit your specimen.

Colours indicate lineage
Hydrocharitaceae
Cymodoceaceae
Posidoniaceae Ruppiaceae
Zosteraceae
Potamogetonaceae

Cylindrical

Leaves leathery, difficult to tear
→ *Posidonia ostenfeldii 'complex'*

Leaves not leathery, nor difficult to tear

→ Leaf diameter usually > 1 mm
→ *Syringodium isoetifolium*

→ Leaf diameter < 1 mm, plants fine and threadlike

→ Mature flowers on spiral stalk
→ *Ruppia*

→ Flowers in small clusters at ends of shoots
→ *Lepilaena*

Oval-oblong

→ Obvious vertical stem with 3 or more leaves
→ *Halophila spinulosa*

Leaves with petioles, in pairs

→ Leaf hairs on leaf blade, margins serrated
→ *Halophila decipiens*

→ No leaf hairs, leaf margins smooth

→ Flowers attached to rhizome, 3 styles
→ *Halophila ovalis 'complex'*

→ Flowers on vertical stalks, 6 styles
→ *Halophila australis*

Strap-like on long vertical stem

Leaf tips serrated

→ In temperate Western Australia
→ *pachyhizum*

→ Not in temperate Western Australia
→ *Thalassodendron ciliatum*

Leaf tips not serrated

→ Leaf length < 8 cm, leaf cross-veins obscure

→ Leaf sheath overlaps at base only
→ *Amphibolis antarctica*

→ Leaf sheath overlaps along entire sheath
→ *Amphibolis griffithii*

→ Leaf length > 8 cm, leaf cross-veins obvious
→ *Zostera tasmanica 'complex'*

Strap-like no, or very short, vertical stem

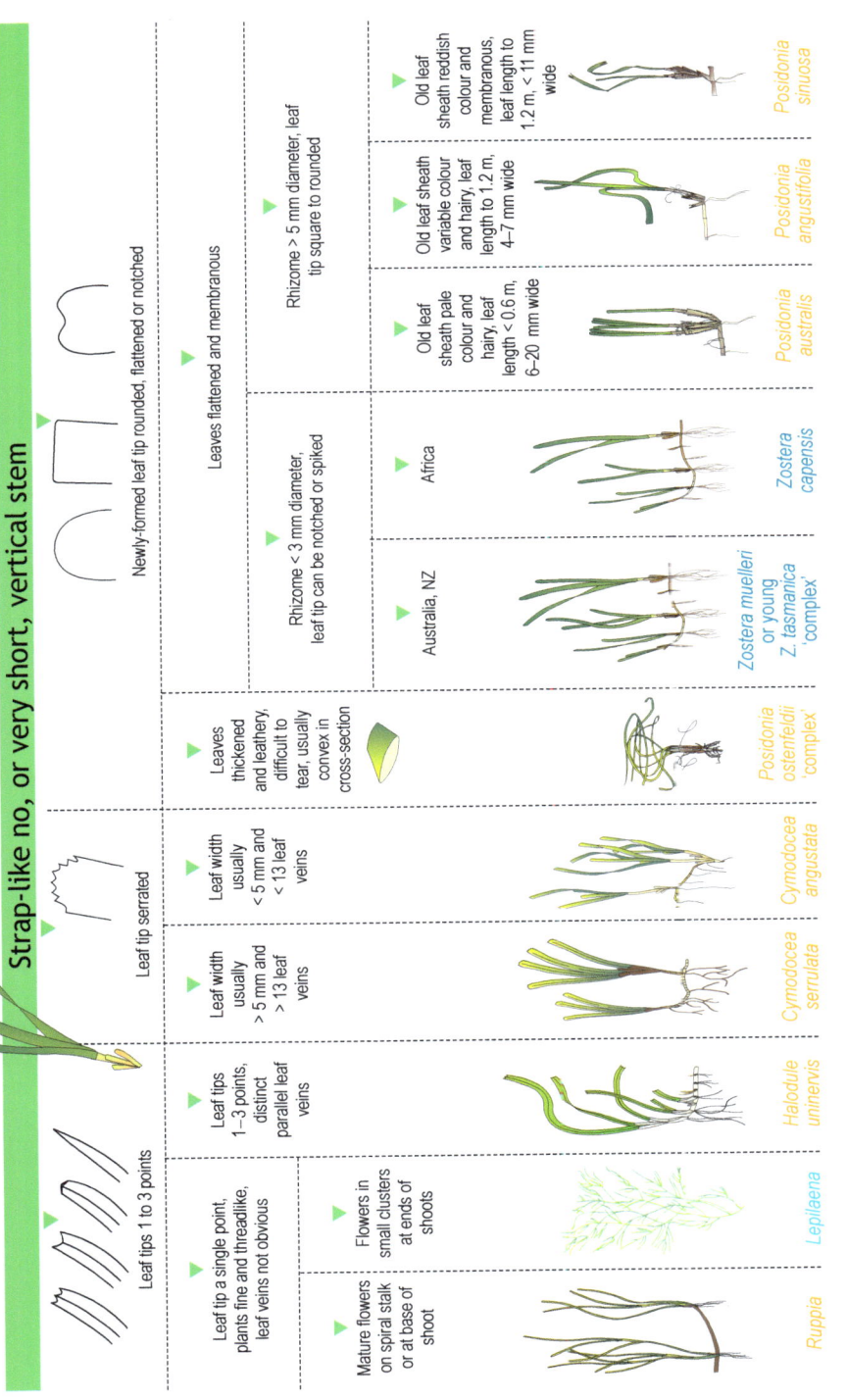

- **Leaf tips 1 to 3 points**
  - Leaf tip a single point, plants fine and threadlike, leaf veins not obvious
    - Flowers in small clusters at ends of shoots — *Lepilaena*
    - Mature flowers on spiral stalk or at base of shoot — *Ruppia*
  - Leaf tips 1–3 points, distinct parallel leaf veins — *Halodule uninervis*
- **Leaf tip serrated**
  - Leaf width usually > 5 mm and > 13 leaf veins — *Cymodocea serrulata*
  - Leaf width usually < 5 mm and < 13 leaf veins — *Cymodocea angustata*
- Leaves thickened and leathery, difficult to tear, usually convex in cross-section — *Posidonia ostenfeldii* 'complex'
- **Newly-formed leaf tip rounded, flattened or notched**
  - Rhizome < 3 mm diameter, leaf tip can be notched or spiked
    - Australia, NZ — *Zostera muelleri* or young *Z. tasmanica* 'complex'
    - Africa — *Zostera capensis*
  - Leaves flattened and membranous
    - Rhizome > 5 mm diameter, leaf tip square to rounded
      - Old leaf sheath pale colour and hairy, leaf length < 0.6 m, 6–20 mm wide — *Posidonia australis*
      - Old leaf sheath variable colour and hairy, leaf length to 1.2 m, 4–7 mm wide — *Posidonia angustifolia*
      - Old leaf sheath reddish colour and membranous, leaf length to 1.2 m, < 11 mm wide — *Posidonia sinuosa*

Species list

Amphibolis antarctica   p40

- ▶ leaf sheath overlapping only at base of sheath
- ▶ leaves generally twisted and 2–5 cm long
- ▶ wiry stem with clusters of 6–8 leaves

Amphibolis griffithii   p42

- ▶ leaf sheath overlapping along the entire length
- ▶ leaves flat and 3–8 cm long
- ▶ wiry stem with clusters of 2–5 leaves

Cymodocea angustata   p44

- ▶ leaves < 5 mm wide, rounded, serrated tips, < 13 veins
- ▶ leaves arise from short vertical stem with leaf scars & no persistent leaf sheaths
- ▶ single unbranched root at node

Cymodocea serrulata   p46

- ▶ leaves > 5 mm wide, with rounded, serrated tips, > 13 veins
- ▶ leaves arise from a short vertical stem, with sheath forming 'V' on one side, leaving open leaf scars on stem
- ▶ 2–3 branched roots at node

Halodule uninervis   p48

- ▶ pointed leaf tip (one to three points) with central leaf vein to the tip
- ▶ distinctive vertical leaf veins and obscure cross-veins
- ▶ pale coloured rhizome, the nodes encircled by dark fibres

Syringodium isoetifolium   p50

- ▶ cylindrical leaves containing air cavities
- ▶ forms small patches within mixed meadows
- ▶ tropical Indo-West Pacific, Australia—south to Geographe Bay, Eastern Africa—south to Mozambique

Thalassodendron ciliatum   p52

- ▶ leaves > 10 cm long, curving, with rounded, serrated tips on wiry vertical stem
- ▶ tropical Indo-West Pacific, Eastern Africa—south to South Africa

Thalassodendron pachyrhizum   p54

- ▶ leaves > 7 cm long, curving, with rounded, serrated tips on wiry vertical stem
- ▶ endemic to southern Australia

Posidonia angustifolia   p58

- ▶ leaves typically 4–7 mm wide, with tattered tips and fibrous old sheaths
- ▶ intact leaf tips oblique
- ▶ inflorescence within canopy

Posidonia australis   p60

- ▶ leaves typically > 6 mm up to 20 mm wide, with rounded leaf tips, up to 60 cm long
- ▶ fibrous old leaf sheaths
- ▶ inflorescence above canopy

Posidonia ostenfeldii 'complex'   p62

- ▶ thickened leaves with tough, leathery texture and distinct lengthways curl
- ▶ distinct horizontal veins at base of leaves
- ▶ cross-sectional leaf shape variable—cylindrical to biconvex to flattened—but always thick

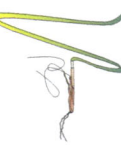

Posidonia sinuosa   p66

▶ leaves 4–11 mm wide,
up to 1.2 m long & with
membranous old leaf sheaths
▶ rhizome > 5 mm diameter
▶ inflorescence within canopy

Ruppia   p70

▶ mature inflorescence on
spiralled stalk or at base
▶ fine cylindrical leaves with
pointed leaf tips, < 1 mm wide
▶ widespread through
temperate and tropical regions

Halophila australis   p76

▶ female flower has 6 styles
and is held on an erect stalk,
not at the base of the rhizome
▶ leaves with smooth edges
▶ leaf cross-veins at 45°
▶ no hairs, leaves in pairs

Halophila decipiens   p78

▶ leaves with minute hairs on
both sides of blade and minute
serrations along the edge
▶ widely distributed
throughout temperate and
tropical habitats
▶ paddle-shaped leaves, the
length longer than the width

Halophila ovalis 'complex'   p80

▶ female flowers with 3
styles attached to rhizome
▶ no hairs, leaves in pairs
▶ leaves with smooth edges
▶ widely distributed
throughout temperate and
tropical regions

Halophila spinu'osa   p84

▶ pairs of leaves arising on
opposite sides of a central
vertical stem, in a single plane
▶ leaves with minute
serrations on edges and a
small one-sided 'fo d' at
the base
▶ central south Asian tropical
distribution, Australia—south
to Jurien and Moreton bays

Zostera capensis   p88

▶ leaves with visible cross-
veins, arise directly from
rhizome with membranous
older leaf sheaths
▶ rhizome < 3 mm diameter,
usually dark brown or yellow
▶ endemic to eastern Africa

Zostera muelleri   p90

▶ leaves with visible cross-
veins, arise directly from
rhizome with membranous
older leaf sheath
▶ rhizome < 3 mm, usually
dark brown or yellow
▶ in Australia and
New Zealand

Zostera tasmanica 'complex'   p92

▶ persistent sheaths present
on erect stem
▶ Leaf < 3 mm wide, > 8 cm
long with obvious cross-veins
▶ found throughout
temperate Australia and Chile

Lepilaena   p98

▶ flowers borne in clusters at
the end of the shoot
▶ fine, cylindrical leaves,
▶ < 1 mm wide with pointed
tips widespread

Cymodoceaceae

The family Cymodoceaceae includes only marine species and is closely related to two other seagrass families, the Ruppiaceae and the Posidoniaceae. There are five genera in this family: *Amphibolis*, *Cymodocea*, *Thalassodendron*, *Syringodium* and *Halodule*. *Amphibolis* is the only predominantly temperate genus, although the species *A. antarctica* extends into warmer waters along the Western Australian coast. Within the genus *Thalassodendron*, *T. pachyrhizum* is restricted to mostly temperate waters. The remaining genera have a primarily tropical distribution but extend into cooler waters, such as *T. ciliatum* in south-eastern Africa and *C. serrulata* in Western Australia.

Genera of the Cymodoceaceae contain a considerable diversity of plant forms, with wiry, vertical stems in *Thalassodendron*, tiny, fine leaves in *Halodule* and air-filled, tube-like shoots in *Syringodium*. The species diversity within genera mostly correlates with major oceanic disjunctions of the Indo-West Pacific and the Caribbean/Tropical Atlantic. The diversity of plant form is matched by diverse habitat occupation. For example, the wiry *Thalassodendron* occupies bare reef and rock, often with strong wave action, whereas the fine-leaved *Halodule* can be found in muddy or sandy intertidal areas.

Some of the earliest unambiguous seagrass fossils are members of this family. In particular, the fossils of the Middle Eocene (35–45 million years ago) Avon Park Formation in Florida USA, contain representatives of *Cymodocea* and *Thalassodendron*. Very little morphological differentiation is observed between these ancient fossils and their modern day counterparts.

Opposite top: *Thalassodendron pachyrhizum*.
Opposite bottom: *Amphibolis antarctica*.

Amphibolis antarctica (Labill.) Sond. & Asch. ex Asch.

CR

Amphibolis antarctica is endemic to southern Australia where it forms extensive meadows in high-energy sandy environments but can also be found on reefs and rocky walls. The structurally complex canopy reduces water motion, assists with the accumulation of organic matter in sediments and supports one of the most biologically diverse and productive of all seagrass habitats. Its detached leaves and stems are a common component of the marine 'wrack' which washes onto beaches and supports beach food webs. In Shark Bay, this species is an important food for dugongs during the winter months, when cool water temperatures force them out of their preferred foraging habitats.

DK

Curled leaves arranged in clusters

Taxonomy and morphology

A. antarctica is distinguished from the other species in this genus, *A. griffithii*, by its more tightly clustered, shorter leaves (2–5 cm) that twist slightly, and by leaf sheaths that overlap only at the base, not along the entire sheath. The long, wiry stem is clearly marked by the scars of old leaves that have been shed. It may be confused with *Thalassodendron pachyrhizum*, which has fewer leaf clusters and longer leaves.

Cymodoceaceae

Amphibolis antarctica

Reproduction

A. antarctica has cryptic male and female flowers on separate plants, up to 20 per stem. Male flowers produce the longest pollen grains on earth (forming filaments up to 5 mm long) and are transported by currents to female flowers for pollination. Fruits rapidly develop into seedlings on the female plant (i.e. viviparous). A bristled comb at the base of the seedling helps it attach after release. Seedlings are often seen washed up on the beach in autumn and winter.

Male flower

JV

Release of clumps of pollen

JV

Immature seedling

KM

Seed

Habitats

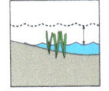
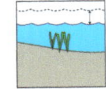

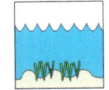

Flowering
Fruiting

J F M A M J J A S O N D

Distribution

LC

Key characteristics

- Long wiry stem with clusters of 6–8 leaves
- Viviparous seedlings with distinctive comb at base
- Leaf apex truncate or semi-circular with two lateral teeth
- Leaves shorter than *A. griffithii* (2–5 vs. 3–8 cm)
- Endemic to southern Australia

Distinctive feature

- Leaf sheath overlapping only at base

Amphibolis griffithii

(J.M.Black) Hartog

Amphibolis griffithii has a more restricted distribution, is less common on reefs and tends to occur in slightly deeper water than *A. antarctica*, on average down to 12 m, but it has been recorded as deep as 48 m. It can form monospecific meadows or it grows mixed with many other seagrass species. The tough stems can persist for over two years and provide an ideal habitat for many epiphytic organisms. The structurally complex habitat supports one of the most diverse and productive seagrass communities. Its detached leaves commonly accumulate on beaches, providing important habitat for fauna.

Flat leaves arranged in clusters

Taxonomy and morphology

The long wiry stem of a mature *A. griffithii* plant is marked by the scars of old leaves that have been shed. The leaves form clusters at the end of the often-branched vertical stem. One of two species in the genus, *A. griffithii* has longer leaves (3–8 cm) that form looser clusters than *A. antarctica*. The leaves rarely twist and the sheath overlaps along its entire length. This species may be confused with *Thalassodendron pachyrhizum*, which has fewer leaf clusters and longer leaves.

Cymodoceaea

Reproduction

Like *A. antarctica*, *A. griffithii* has small cryptic male and female flowers on separate plants with long, filamentous pollen. Flowers develop over winter and spring and pollen is released into the water where currents transport it to female flowers. After pollination, new seedlings develop while still joined to the mother and can remain attached for up to 5 months. Seedlings can float in the water for days to weeks before settling and attaching to the substrate with the comb.

Male flower

Female flower

Seedlings

KM

Seed

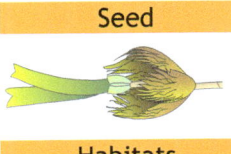

Habitats

Flowering / Fruiting

J	F	M	A	M	J	J	A	S	O	N	D

Distribution

LC

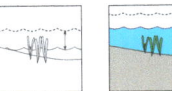

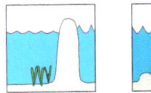 

Key characteristics

▶ Wiry stem with cluster of leaves at end, typically 2–5 leaves per clusters
▶ Viviparous seedlings with distinctive comb at base
▶ Leaf apex deeply notched with two lateral teeth
▶ Leaves flat not twisted and longer than *A. antarctica* (3–8 vs. 2–5 cm)
▶ Endemic southern Australia

Distinctive feature

▶ Leaf sheath overlapping along entire length

Cymodocea angustata

Ostenf.

JS

Cymodocea angustata is uncommon and restricted to the west coast of Australia, south to Shark Bay, with primarily a tropical distribution. Little is known about this species as it has rarely been studied. It has been found with other species of *Cymodocea*, *Halophila ovalis*, *Halodule uninervis* and *Syringodium isoetifolium*. The ecology and habitat requirements may be similar to other *Cymodocea* species. It has been found on all sediment types, from fine muds through to rock on reef flats, typically in shallow water (2–3 m).

JS

C. angustata in Shark Bay

Taxonomy and morphology

C. angustata is uncommon compared to the more widespread *C. serrulata* which is similar in overall morphology. These two species do co-occur. *C. angustata* may be distinguished by its narrower leaves (< 5 mm) with fewer leaf veins (< 13) and only one unbranched root at each node. *Cymodocea* species may also be confused with smaller plants of *Posidonia australis* in Shark Bay, however, *Posidonia* has persistent leaf sheaths, thicker rhizome, typically wider (6–20 vs. 3–6 mm) leaves and smooth leaf tips.

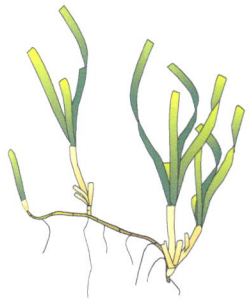

Reproduction

Very little is known about flowering in *C. angustata*. Male and female flowers are on separate plants, but male flowers have rarely been observed. Female flowers emerge from the leaf sheath with 2 long stigmas and have been recorded between September and November. The fruit is around 6 mm long and develops at the base of the plant on a stalk, 4–5 mm long.

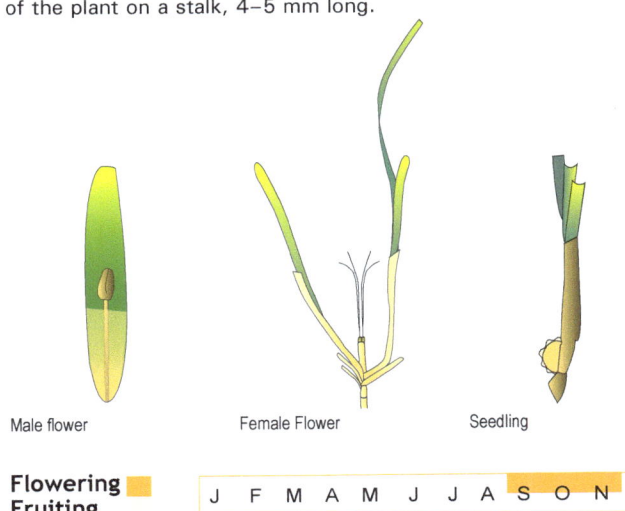

Male flower   Female Flower   Seedling

Seed

Habitats

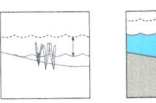 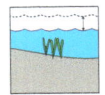
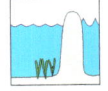 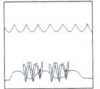
 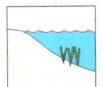

Flowering
Fruiting

| J | F | M | A | M | J | J | A | S | O | N | D |

Unknown in this region

Distribution

LC

Key characteristics

▶ Leaves arise from short vertical stem with rounded serrated leaf tips
▶ A single unbranched root at each node
▶ Short vertical stem with obvious leaf scars
▶ No persistent leaf sheaths
▶ Endemic to NW Australia, southern limit Shark Bay

Distinctive feature

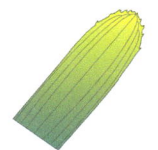

▶ Leaf width < 5 mm, < 13 parallel leaf veins

Cymodocea serrulata

(R.Br.) Asch. & Magnus

This species is very widespread with a primarily tropical distribution, common throughout the Indo-West Pacific, extending southwards to Shark Bay in Australia and to southern Mozambique in Africa. *Cymodocea serrulata* is usually the dominant species in mixed-species communities on muddy reef tops or, less commonly, silty depositional sands. It is considered to be relatively tolerant of burial and adapted to more turbid environments, typically growing in 3–4 m of water. In meadows, *C. serrulata* can be confused with *Posidonia australis* and the wide-leaved forms of *Halodule uninervis*, while loose leaves can be confused with *Thalassodendron ciliatum*, so it is important to look for the key characteristics to identify each species.

Wider leaves of *C. serrulata*

Taxonomy and morphology

In overall morphology *C. serrulata* may be confused with *C. angustata*. The two species co-occur but *C. serrulata* can be distinguished by its wider leaves (> 5 mm) with more leaf veins (> 13). *C. serrulata* is relatively common in warmer waters and subtropical locations. *Cymodocea* species may also be confused with smaller plants of *Posidonia australis* in Shark Bay, which may be differentiated by their persistent leaf sheath, thicker rhizome, typically wider (6–20 vs. 4–10 mm) leaves and smooth leaf tip.

Cymodoceaceae

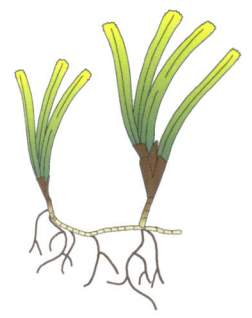

Cymodocea serrulata

Reproduction

Male and female flowers form on separate plants in *Cymodocea serrulata*. The male flower is stalked, having stamens emerging from the leaf sheath and filamentous pollen without an exine. Female flowers form at the base of the leaf with deeply forked styles. The fruit contains nut-like seeds which are dark coloured with three prominent ridges up to 10 mm long and can form a seed bank. The timing of flowering in the southern temperate region is unknown.

Female flower   KV

Male flower   KV

Seedling   MW

Seed

Habitats

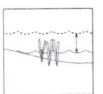 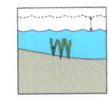

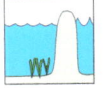

**Flowering
Fruiting**

| J | F | M | A | M | J | J | A | S | O | N | D |

Unknown in this region

Distribution

LC

Key characteristics

▶ Leaves arise from a short vertical stem with rounded, serrated tips
▶ 2–3 branched roots at a node
▶ Leaf sheath flat and forming 'V' on one side, leaving open leaf scars on stem
▶ Australia, south to Shark Bay and eastern Africa, south to Mozambique

Distinctive feature

▶ Leaf width > 5 mm, > 13 parallel leaf veins

Halodule uninervis
(includes *H. pinifolia*)

(Forssk.) Asch.

KM

Halodule uninervis has primarily a tropical distribution and is widely distributed throughout the Indo-West Pacific, but it does extend into the temperate zone at the southern limits of its distribution on the west coast of Australia and in southern Mozambique in Africa. It can occupy a wide range of habitats, from the muddy intertidal to reef tops and coarse sand, and is tolerant of large fluctuations in salinity. It rapidly colonises from both seed and vegetative growth, forming sparse meadows. It plays an important role in stabilising sediments in disturbed areas through the spread of its intertwining rhizome mat and fibrous roots.

KM

Rhizome and shoots of *H. uninervis*

Taxonomy and morphology

Halodule uninervis is readily differentiated from other temperate seagrasses by the presence of 1–3 points on the leaf tips and a clearly seen central vein on the leaves. This plant may be confused with *Ruppia* spp. in overall morphology, however, leaves of *Ruppia* are generally much finer (< 1 mm) and always have a single point to the ends of the leaf tips.

Reproduction

Halodule uninervis has separate male and female plants with obscure flowers, forming in the base of the leaf sheath, usually buried beneath the sediment, only emerging for a short time when fully mature. Flowering in *Halodule uninervis* is common in the tropics but is unknown in this region, though it is likely to occur. The hard, dark seeds, one or two per shoot, form at the base of the shoot and remain attached to the rhizome for weeks to months, and can also form a seed bank.

Seed

Habitats

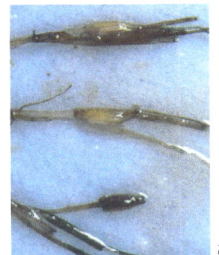

Male flowers

Developing fruit

Seeds

 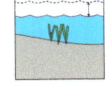
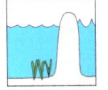 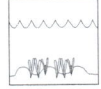

Flowering
Fruiting

| J | F | M | A | M | J | J | A | S | O | N | D |

Unknown in this region

Distribution

LC

Key characteristics

▶ Relatively small plant with leaves that can arise from a vertical stem
▶ Pale coloured rhizome, the nodes encircled by dark fibres
▶ Distinctive vertical leaf veins and obscure cross-veins
▶ Found throughout the Indo-West Pacific, south to Jurien Bay and Eastern Africa, south to Mozambique

Distinctive feature

▶ Leaf tip 1–3 points, distinct central leaf vein

Syringodium isoetifolium

(Asch.) Dandy

Syringodium isoetifolium is widely distributed in both temperate and tropical habitats. On the west coast of Australia, this species grows down to 33°S due to the influence of the warm water Leeuwin Current and in Africa it is found in southern Mozambique. In temperate regions, *S. isoetifolium* is generally a patchy, sub-dominant member of mixed-species meadows. It can colonise disturbed areas, though its presence can be strongly seasonal, with higher biomass in summer. It is grazed by some sea urchins (diadematids) but recovers relatively quickly since its growing meristem is typically buried below the sediment, protected from grazing.

Branched, cylindrical leaves

Taxonomy and morphology

The leaves of *Syringodium* are unusual as they are hollow and full of air; when they break off they float. The leaves are cylindrical rather than flat and so superficially can be confused with *Ruppia* spp. or the terete form of *Posidonia ostenfeldii* (though unlike *P. ostenfeldii* the leaves of *S. isoetifolium* branch). Leaf length can be highly variable in this species, as short as 5 cm to more than 50 cm long in mature plants.

Syringodium isoetifolium

Reproduction

Syringodium isoetifolium has separate male and female plants. The flowers are arranged in a branched structure called a cyme on a vertical stem, which grows to the middle of the canopy. The male flower has a pair of pale pink or red anthers which produce filamentous pollen. The female flower has two pale-coloured, split, prong-like stigmas. The mature seeds are usually dark-coloured, and beaked with a smooth outer covering. The cyme, containing mature seeds, can break off and float considerable distances, aiding dispersal, otherwise the seeds sink.

Flowering structure – cyme   KM

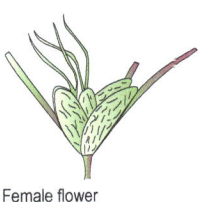

Male flower

Female flower

Seed

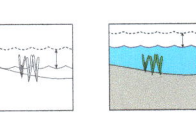

Habitats

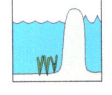 

Flowering / Fruiting

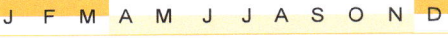

J F M A M J J A S O N D

Distribution

LC

Key characteristics

▶ Shoots may branch and be more than 50 cm long, usually standing erect
▶ Flowers borne on complex structure called a 'cyme'
▶ Forms small patches within mixed meadows
▶ Found throughout the tropical Indo-West Pacific, south to Geographe Bay and Eastern Africa south to southern Mozambique

Distinctive feature

▶ Cylindrical leaves containing air cavities

Thalassodendron ciliatum

(Forssk.) Hartog

This species has a primarily tropical distribution but is also found in temperate waters of northern South Africa. The tough rhizomes and roots of *Thalassodendron ciliatum* provide a strong grasping anchor that makes it well suited to rocky substrates with a high degree of water movement, such as reefs. Since many other seagrasses cannot survive in these habitats, it typically forms single-species meadows. *T. ciliatum* is often heavily grazed by sea urchins, but re-sprouts from the rhizome. The other species from the same genus, *T. pachyrhizum*, occurs only in temperate waters.

Leaf clusters at end of wiry stem

Taxonomy and morphology

The leaves of this seagrass are found in clusters at the end of a tough wiry stem. The leaves emerge from a distinct leaf sheath, at the base of the oldest leaf, and the leaf tip is serrated. The vertical stem is clearly marked by the scars of old leaves that have been shed. While this species may be confused with *T. pachyrhizum* they do not co-occur in temperate waters, *T. pachyrhizum* is found in Australia and *T. ciliatum* in Africa.

Cymodoceaceae

Reproduction

Thalassodendron ciliatum has separate male and female plants. Two paired flowers form on specialised shoots. In the male flower, the anthers are 6–7 mm long, yellow-red in colour and form filamentous pollen. The female flower has a pale-coloured, long (20 mm) forked stigma. Like *Amphibolis*, the seedlings of *T. ciliatum* develop on the mother plant (vivipary) but lack the bristle comb of *Amphibolis*. The bract from the female flower swells to form a buoyant leaf so that when the seedling is released it can be dispersed on ocean currents.

Male

Female

Seedling

Seed

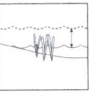

Habitats

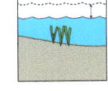
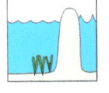
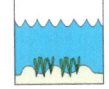

Flowering Fruiting

J	F	M	A	M	J	J	A	S	O	N	D

Unknown in this region

Distribution

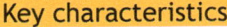

LC

Key characteristics

▶ Leaves greater than 10 cm long and curving with a rounded, serrated leaf tip

▶ Rhizome thick and hardy with tough roots allowing attachment to hard substratum

▶ Viviparous seedlings remain on maternal shoot until at least 3 cm in length

▶ Found throughout the Tropical Indo-West Pacific and eastern Africa south to northern South Africa

Distinctive feature

▶ Rounded, serrated leaf tips on vertical stem, Africa

Thalassodendron pachyrhizum

Hartog

KV

Thalassodendron pachyrhizum is endemic to south-western Australia, with a relatively restricted distribution, from Geraldton to Bremer Bay. It forms meadows in sandy substratum but the high attachment strength of its roots and rhizome also allows it to occur on reef surfaces and tolerate high-energy habitats. It has been observed growing to depths of 40 m in clear waters. It forms small patches or clumps, and can grow among both species of *Amphibolis*, which are superficially similar. Little is known about the ecology of this species.

KV

Tough, blackened rhizome with wiry stem

Taxonomy and morphology

The leaves of this seagrass are found in clusters at the end of a tough wiry stem, emerging from a distinct leaf sheath. A scar is left on the stem when leaves are shed. The species may be confused with *T. ciliatum*, but in temperate regions they do not co-occur; *T. pachyrhizum* occurs in Australia, *T. ciliatum* in Africa. The overall morphology of *Thalassodendron* resembles *Amphibolis*, however, it may be readily distinguished by leaves of more than 7 cm in length, and leaf tips that are rounded with serrations.

Thalassodendron pachyrhizum

Reproduction

Thalassodendron pachyrhizum has separate male and female plants with similar reproductive features to *T. ciliatum*. Around 10% of stems produce flowers and flowering is estimated to occur every 3–4 years. The seedling develops on the mother plant (vivipary) and the outer bract of the female flower swells to produce a buoyant structure, aiding dispersal away from the mother plant. The swollen bract tends to be at least two times larger than the other bracts in *T. pachyrhizum* distinguishing it from *T. ciliatum* where the bracts are of similar size.

Male

Female

Detached seedling on beach

KM

Seed

Habitats

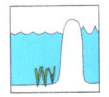

Flowering
Fruiting

J F M A M J J A S O N D

Distribution

LC

Key characteristics

▶ Leaves greater than 7 cm long and curving with a rounded, serrated leaf tip
▶ Tough, woody rhizome
▶ Viviparous seedlings remain on maternal shoot until at least 3 cm in length
▶ Up to 3 leaf clusters on wiry vertical stem
▶ Endemic to southern Australia

Distinctive feature

▶ Rounded, serrated leaf tips on vertical stem, Australia

Posidoniaceae

Members of the genus *Posidonia* have been recognised as belonging to their own family since Cronquist's revision of flowering plant families in 1981. Prior to this, including in den Hartog's 1970 *The sea-grasses of the world*, *Posidonia* had been considered to be a part of the Potamogetonaceae along with the majority of other seagrass genera outside of the Hydrocharitaceae. Here we recognise the family Posidoniaceae, broadly following the work of Cronquist.

Recent molecular analyses demonstrate that the Cymodoceaceae, Ruppiaceae and Posidoniaceae represent an independent evolutionary lineage. The Posidoniaceae are the sister family to the Ruppiaceae and the Cymodoceaceae contains only one genus and up to nine species. Only one species occurs outside of Australia, *Posidonia oceanica*, which is widespread but restricted to the Mediterranean Sea. The Australian species occur as two clearly distinguishable groups: the *Posidonia australis* 'complex', comprising *P. australis, P. sinuosa* and *P. angustifolia*—these species occur in relatively sheltered locations; and the *Posidonia ostenfeldii* 'complex' comprising *P. ostenfeldii, P. coriacea, P. robertsoniae, P. denhartogii* and *P. kirkmanii*—all of which are found in more exposed, higher-energy locations. The morphology of this group follows a basic form with strap-like leaves and a buried meristem directly attached to buried rhizomes. Members of the *P. ostenfeldii* 'complex' have leathery, tough leaves with deep rhizomes, helping them cope with the high-energy environments they occupy.

Recent analysis of the currently recognised species in these two groups have demonstrated that members of the *P. australis* 'complex' are reasonably distinct taxa although *P. angustifolia* may be of hybrid origin—we treat all three separately in this guide. In contrast, there is little support for continuing to recognise the different species in the *P. ostenfeldii* 'complex' despite a lot of variability in leaf width—we treat these as a single taxon although we outline the differences among the currently recognised species.

Opposite top: *Posidonia australis.*
Opposite bottom: *Posidonia sinuosa.*

Posidonia angustifolia

Cambridge & J.Kuo

KM

Posidonia angustifolia is endemic to southern Australia, and one of three species that comprise the *P. australis* 'complex'. Like other members of the complex, it has horizontal rhizome growth and forms dense, monospecific or mixed-species meadows, typically in sheltered waters. It has been recorded to depths of 44 m. In higher-energy environments it can co-occur with members of the *P. ostenfeldii* 'complex'. The meadows are highly productive and though they are rarely grazed they provide an important habitat for a diverse array of fish and invertebrate fauna, including abundant filter-feeders capable of filtering the overlying water column on a daily basis. The species is slow to recover (years to decades) following disturbance.

GK

Patches of *P. angustifolia*

Taxonomy and morphology

The identification of this species can be confusing as it shares some morphological features with both *P. australis* and *P. sinuosa. Posidonia angustifolia* is often misidentified as *P. australis* but has narrower and longer leaves. The rhizome is laterally compressed and may be distinguished from *P. sinuosa* by the presence of a persistent, hairy leaf sheath, as opposed to a membranous sheath in *P. sinuosa*. The long leaves are readily torn and as a result the leaf tips are often tattered and irregular.

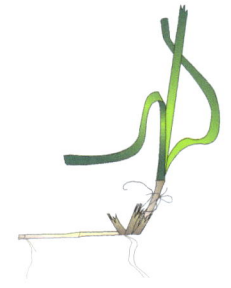

Reproduction

The flowers of *Posidonia* are different to most other seagrass genera as they are hermaphroditic (male and female parts being within the same flower). The flowers are arranged on a spike contained within the canopy. The fruits are green in colour and called drupes as they have a grey-green skin that surrounds fleshy, green tissue and then the seed. The seeds have no dormancy, germinating soon after release from the plant. Little is known about flowering intensity for *P. angustifolia*.

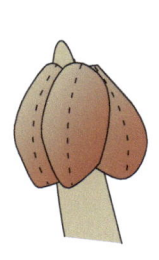

Male flower parts

Female flower parts

Inflorescence

Seed

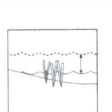

Habitats

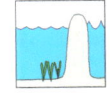
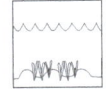

Flowering
Fruiting

J F M A M J J A S O N D

Distribution

LC

Key characteristics

▶ Leaves 4–7 mm wide
▶ Intact leaf tips oblique
▶ Rhizome usually laterally compressed
▶ Inflorescence within canopy
▶ Endemic to southern Australia

Distinctive feature

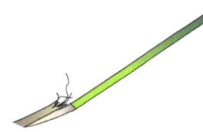

▶ Leaves < 7 mm wide, tips frayed; old sheath hairy

Posidonia australis

Hook.f.

JG

Posidonia australis is a meadow-forming seagrass distributed throughout southern Australia. On the east coast, it is confined to estuaries and lagoons but on the west coast it occurs in a range of habitats, including sheltered coastal embayments. The dense meadows are often monospecific, although it is frequently found with other *Posidonia* species. It typically dominates *P. sinuosa* in very shallow and more disturbed areas, where shorter and wider leaves are resistant to mechanical damage caused by water movement. In Shark Bay, it tolerates salinities up to 1.5 times that of seawater. It is rarely grazed directly (though some fish, principally wrasses, and a portunid crab do consume leaves) but provides habitat for a diverse and highly productive community of fauna, including commercial fishery species.

Edge of small patch

KM

Taxonomy and morphology

This species was split into three in 1979 with the two new species, *P. angustifolia* and *P. sinuosa. Posidonia australis* has wider but shorter leaves, distinguishing it from *P. angustifolia*, and hairy (fibrous) old leaf sheaths, distinguishing it from *P. sinuosa*. The leaves are often slightly curved and leaf tips are usually intact. As this species is very common in temperate Australia, it is likely that you will encounter it.

Posidoniaceae

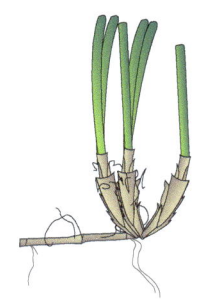

Posidonia australis

Reproduction

The flowers of *Posidonia australis* are hermaphroditic. They are found on a long stalk (20–30 cm) that extends above the canopy with 3–5 spikes at the end of the stalk and 4–6 flowers per spike. Flowers begin to develop in early winter and reach maturity after about four months. Filamentous pollen is released during spring, and successively from flowers at the top of the spike downwards. The green drupe fruit develops on the plant, up to 500 m⁻², and are released in summer when large numbers can be found as drift; they are also eaten by a variety of crabs within the meadows.

Inflorescence

Inflorescence underwater

Fruit

Seed

Habitats

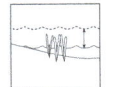 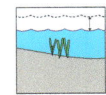

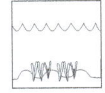

Flowering Fruiting

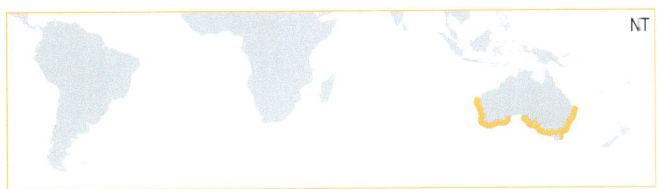

J F M A M J J A S O N D

Distribution

NT

Key characteristics

▶ Leaves 6–20 mm wide and up to 60 cm long
▶ Rhizome usually laterally compressed
▶ Inflorescence above canopy
▶ Endemic to southern Australia

Distinctive feature

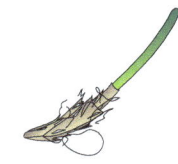

▶ Leaves > 10 mm, tips rounded; old sheath hairy

Posidonia ostenfeldii 'complex'

KM

Five closely related species of *Posidonia* are grouped together in the *Posidonia ostenfeldii* 'complex'. All are characterised by common leaf and rhizome features that permit them to survive in high-energy, open ocean environments. They all have long, tough, leathery leaves arising from deeply buried rhizomes, which, unlike any other seagrass species, grow in a vertical direction and branch such that the leaves emerge from the sediment in small, almost linear patches, forming a sparse and patchy meadow. The patchy nature of the meadows supports a less diverse community than the *P. australis* 'complex' meadows. These species could be confused with other *Posidonia* or with *Syringodium isoetifolium*, so attention should be paid to the distinguishing features when identifying specimens.

KM

Leaves with lengthwise curl

Taxonomy and morphology

There is large morphological variation in this group of species making them difficult to identify. As a result, we treat this group as a species 'complex' including *P. ostenfeldii*, *P. coriacea*, *P. denhartogii*, *P. kirkmanii* and *P. robertsoniae* (see following page). The form of these plants is robust, with tough, thickened leaves, vertical rhizomes and deep roots that pull the plant into the sediment. Leaf width, length, thickness and shape vary considerably.

Reproduction

The members of the *P. ostenfeldii* 'complex' also have hermaphroditic flowers and fruits in spikes on a stalk (20–40 cm) that terminates within the canopy. Seed production is less than in other *Posidonia* species, up to 15 m^{-2}, although this is enough to maintain the population. As in other *Posidonia*, the fruit is buoyant, aiding dispersal to other areas where germination can occur.

Flowering spike within canopy

Fruit on detached spike

Seedlings

Seed

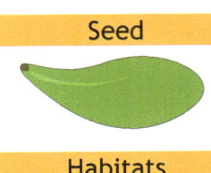

Habitats

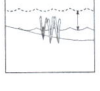
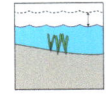
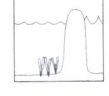
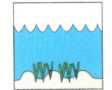

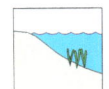

Flowering Fruiting

J F M A M J J A S O N D

Distribution

LC

Key characteristics

▶ Deep contractile roots
▶ Leaves up to 180 cm long, 12 mm wide and difficult to tear crosswise
▶ Cross-sectional leaf shape highly variable—cylindrical to biconvex to flattened but always thick
▶ Leaves with distinct lengthways curl
▶ Distinct horizontal veins at base of leaves

Distinctive feature

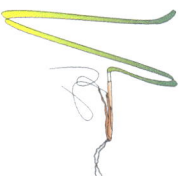

▶ Tough, leathery, thickened leaves

Posidonia ostenfeldii 'complex'

Five species have been described in the *Posidonia ostenfeldii* 'complex' but there is disagreement as to the number of species. Many of the key characteristics that have been used to define the species are not discrete but overlap. This makes it difficult to distinguish species in the field. When a large number of samples are examined over a range of sites, there appear to be no clear distinctions, based on their basic morphology, indicating it is one group with a variable morphology. This concept is supported by a number of studies that have examined the genetic differences of representatives of the *P. ostenfeldii* 'complex' and found no clear distinctions, which would support that they are not different species. At this point in time, no formal changes have been made to the taxonomy, however, based on the uncertainty described above, we present this group of species as a 'complex'. Research is continuing in this area to understand species concepts within this 'complex'. We include descriptions and key characteristics of the currently described species in the *P. ostenfeldii* 'complex'. These species occupy the same habitat types and their adaptations to survive in high-energy environments clearly represent a divergence from the remainder of the *Posidonia* species.

Posidonia coriacea   Cambridge & J.Kuo

Key characteristics
- ▶ Leaves biconvex in cross-section
- ▶ Leaves 2.5–7 mm wide with 2–3 leaves per shoot
- ▶ Leaf surface rough (coriaceous)

Posidonia denhartogii   J.Kuo & Cambridge

Key characteristics
- ▶ Leaves biconvex in cross-section
- ▶ Leaves 1–2 mm wide with 2–3 leaves per shoot

Posidonia kirkmanii   J.Kuo & Cambridge

Key characteristics
- ▶ Leaves biconvex in cross-section
- ▶ Leaves 6–12 mm wide with 2–3 leaves per shoot

Posidonia ostenfeldii   Hartog

Key characteristics
- ▶ Leaves circular in cross-section
- ▶ Leaves 1–1.5 mm wide with 1–3 leaves per shoot

Posidonia robertsoniae   J.Kuo & Cambridge

Key characteristics
- ▶ Leaves flattened in cross-section
- ▶ Leaves 2–4.5 mm wide with 1–2 leaves per shoot

Posidonia sinuosa

Cambridge & J.Kuo

KM

Posidonia sinuosa differs from the other members of the *P. australis* 'complex' in both its shoot and meadow characteristics. It has high shoot densities up to 1500 m^{-2}, often in rows 20–30 cm apart which produces a characteristic 'wind-row' appearance when the meadows are viewed from above. It is usually found in sheltered embayments along with *P. australis* but typically dominates in slightly deeper or more sheltered water. The meadows are important habitat for a diverse community and are an important source of detritus to beaches. Like *P. australis*, the species is slow to recover from disturbance and there have been significant losses throughout its range due to shading associated with nutrient pollution and sediment deposition.

CR

Twisted leaves of *P. sinuosa*

Taxonomy and morphology

This species may be confused with *P. angustifolia* and *P. australis*, however, *P. sinuosa* has persistent membranous leaf sheaths, not hairy (fibrous) ones, and a denser erect growth form. The leaves are long, up to 1.4 m, curved and paper thin (membranous). Large meadows take on a furrowed appearance as the long leaves form natural rows in the waves. The rhizome is uncompressed in cross-section, appearing circular. The inflorescence is found at the base of the canopy.

Posidonia sinuosa

Reproduction

Like other *Posidonia* species, the flowers of *P. sinuosa* are hermaphroditic, on a short stalk (< 10 cm), with 2–4 spikes at the end of the stalk and 4–6 flowers per spike. The flowers occur below the top of the canopy, so pollen is released inside the canopy, in spring. The location of pollen release as well as a shorter time period that the pollen is viable means that *P. sinuosa* has less pollination success compared to *P. australis*. Despite this, fruits are commonly observed.

Flowering spikes

Flowers within canopy

Fruit

Seed

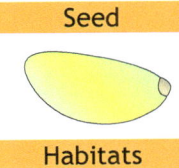

Habitats

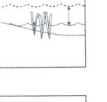
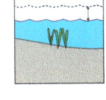
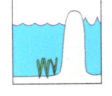
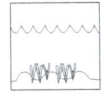

Flowering Fruiting

J F M A M J J A S O N D

Distribution

Key characteristics

▶ Leaves up to 1.4 m long and 4–11 mm wide
▶ Leaves often have distinct concavity in cross-section and twist along length
▶ Inflorescence below canopy
▶ Endemic to southern Australia

Distinctive feature

▶ Membranous old leaf sheaths, rhizome > 5 mm diameter

Ruppiaceae

Found in predominantly temperate environments around the world, members of the genus *Ruppia* are recognised as belonging to their own family, the Ruppiaceae, since Cronquist's revision of flowering plant families in 1981. *Ruppia* is sometimes found in estuaries and ephemeral coastal lagoons but is poorly collected as it is likely to be ephemeral. The Ruppiaceae has been phylogenetically placed within the Cymodoceaceae 'complex' and is most closely related to the Cymodoceaceae and the Posidoniaceae.

Ruppia currently holds an uncomfortable place among the plants we call seagrasses as many of the species and populations of *Ruppia* do not inhabit a strictly marine environment. The species and populations that do not occur in oceanic environments do exist in salt lakes and lagoons, as ephemeral populations heavily reliant on seed recruitment. We consider some members of this genus to be 'seagrasses' as around the world *Ruppia* coexists with other well recognised seagrasses in marine environments.

Ruppia, also known as Widgeon Grass, is an important food resource for waterfowl—ducks, swans and geese—in estuaries and salt lakes. The movement of seeds by these waterfowl most likely contributes to the widespread distribution of the species. *Ruppia* beds are often ephemeral and, when they occur adjacent to more persistent oceanic seagrasses, their presence is sometimes used as evidence of declining ecological health.

Opposite: Fine leaved *Ruppia* plants

Ruppia

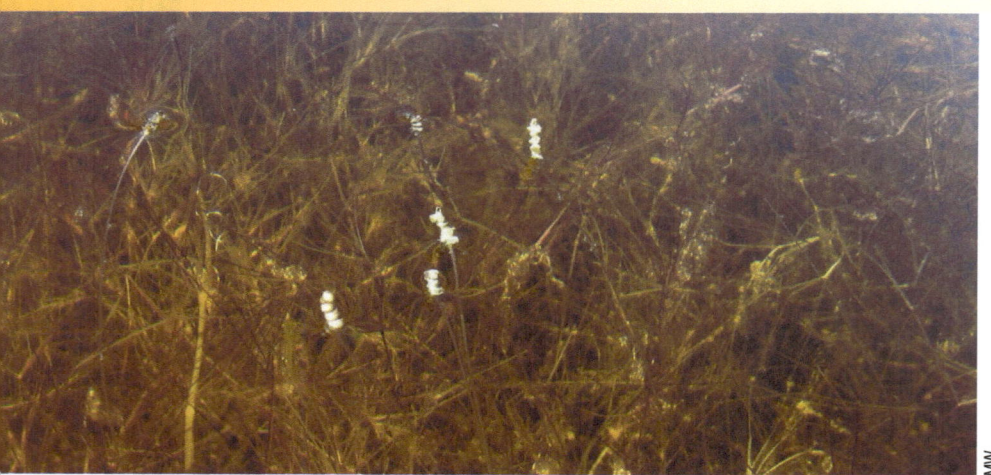

MW

The genus *Ruppia* tolerates a wide range of environmental conditions and, consequently, is widely distributed throughout temperate and tropical regions. It occurs predominantly in semi-fresh or estuarine water, as well as inland saline lakes. It is not always recognised as a seagrass as it occurs in non-marine and marine environments. *Ruppia* plays an important role as a primary producer and provides a complex habitat for other organisms in the absence of other large submerged aquatic plants. It is commonly grazed by waterbirds, particularly swans, and responds quickly to changes in the environment.

KM

Small, fine, branching leaves of *Ruppia*

Taxonomy and morphology

Ruppia is a truly global marine plant genus. The current taxonomy is confounded as different species, described on different continents, may in fact be the same species. Further research is required to resolve this. In the southern temperate region, three species putatively occur in marine habitats. In general *Ruppia* has flattened to cylindrical leaves. While it may be confused with other species with cylindrical leaves, the narrow leaf width, lack of airspaces in the leaves and the production of flowers on spiralled stalks differentiates *Ruppia* from *Syringodium* and *Lepilaena*.

Reproduction

Ruppia forms both male and female flowers on the same plant in a single inflorescence, with male flowers maturing before female flowers. The flowering biology is very different to its most closely related genus, *Halodule*. Inflorescences emerge from a leaf sheath that usually develops on a tight, spiralled stalk that, in some species, sends the flowers to the surface where they are pollinated out of water. Seeds are small (1–2 mm), dark-beaked drupes that are able to persist in sediments for several months to a year.

Inflorescence

KM

Emerging on stalk

KM

Developing seeds

KM

Seed

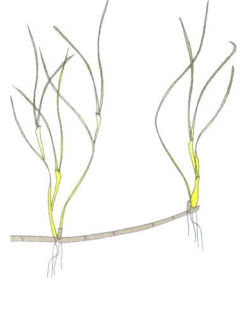

Habitats

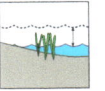 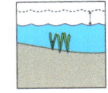

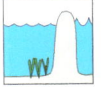 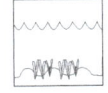

Flowering Fruiting

J F M A M J J A S O N D

Distribution

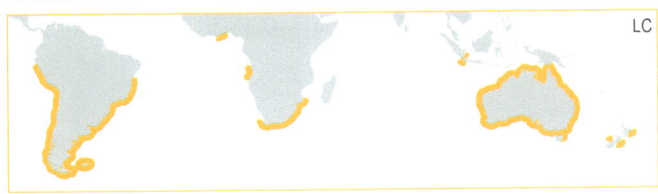

LC

Key characteristics

▶ Fine, cylindrical, leaves
▶ Pointed leaf tips
▶ Leaves < 1 mm wide
▶ Larger plants often much branched
▶ Widespread throughout temperate and tropical regions

Distinctive feature

▶ Mature flower on spiralled stalk or stalk absent

Ruppia species

The species of *Ruppia* included here, are those which are known to grow in estuarine and marine environments. There is some uncertainty about the number of species of *Ruppia* world-wide, between two and ten species having been recognised. Many of the described species do not grow in estuaries or marine habitats, although it is common for them to be found in highly alkaline or saline waters, such as salt lakes and lagoons. Waterfowl feed on *Ruppia* and can transport its seeds when they move between feeding and breeding grounds, on a local scale, or with large-scale migrations. Therefore, like other aquatic plants, a single species can be very widespread, even between continents. The globally distributed *R. maritima* is found in the southern temperate zone in Argentina and Southern Africa, as well as temperate regions of Australia, mostly in estuaries. The marine species of *Ruppia* that grow only in Australia are *R. megacarpa* and *R. tuberosa*. These have been observed growing adjacent to other seagrass species such as *Zostera muelleri*, *Halophila ovalis* and even *Posidonia sinuosa*. Other *Ruppia* species that are found in high salinity environments but have rarely been observed in estuarine environments are *R. polycarpa* and *R. cirrhosa* (which is also known as, or synonymous with, *R. spiralis*).

Ruppia tuberosa J.S.Davis & Toml.

Key characteristics
- ▶ Spiral flowering stalk does not reach water surface
- ▶ Filament holding seeds absent or very short

Ruppia megacarpa R.Mason

Key characteristics
- ▶ Spiral flowering stalk very obvious, reaches water surface
- ▶ Filament holding seeds longer than seed
- ▶ Up to 6 seeds in each inflorescence

Ruppia maritima L.

Key characteristics
- ▶ Spiral flowering stalk very short, does not reach water surface

Hydrocharitaceae

The Hydrocharitaceae is an aquatic plant family found globally in a diverse array of habitats. The majority of species occur in freshwater, with only three of the 17 genera being marine: *Halophila*, *Enhalus* and *Thalassia*. Well-known freshwater genera in this family are *Elodea*, *Hydrocharis* and *Vallisneria*. This is a primarily tropical group of seagrasses but several species of *Halophila* occur in temperate waters.

Among the three marine genera *Halophila* is the most diverse, with 17 species currently recognised. In contrast, *Thalassia* has two species and *Enhalus* only one. There is extreme diversity in morphology among the marine members of this family—from tiny, fragile, round-leaved *Halophila* species, barely 1 cm high, to dense, robust, long-leaved *Enhalus* more than 1 m long.

The relatively high species diversity of *Halophila* may be due to some species having specific habitat preferences and faster growing times than other seagrasses. There is considerable difficulty with the taxonomy of many *Halophila* species due to their morphological plasticity. Here we have adopted a conservative approach to the delimitation of species. One problematic species, *Halophila ovalis*, is described here as a species 'complex', which encompasses a range of previously recognised species and subspecies. *Halophila* species are often commonly known as 'paddleweed' or 'spoon-grass'.

Opposite top: *Halophila ovalis*.
Opposite bottom: *Halophila australis*.

Halophila australis

Doty & B.C.Stone

KM

Halophila australis has a restricted distribution relative to many other *Halophila* species, occurring in south-western and southern Australia. *Halophila australis* often grows in high-energy, disturbed sandy habitats and the specialised flowering shoots that extend the flowers up into the water column, and from which new rhizomes can grow, is likely an adaptation to the sediment movement in these habitats. It can grow in large patches and often with other seagrass species. *H. australis* has not been reported growing intertidally. Like many *Halophila* species there is large morphological variation in leaf shape and size.

FVR

H. australis in Port Phillip Bay, Victoria

Taxonomy and morphology

Halophila australis has a similar appearance to *H. ovalis* until reproductive. Both are small with pairs of oval-shaped leaves attached by petioles to underground rhizomes. When reproductive material forms, the female flower has 6 styles compared to 3 in *H. ovalis*, and new rhizomes can form from the top of reproductive shoots. Previous taxonomy distinguished *H. ovalis* and *H. australis* by leaf size, with *H. australis* having the larger leaves. However, as they have overlapping leaf sizes, this is not a diagnostic feature.

Hydrocharitaceae

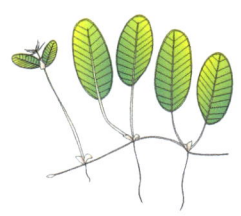

Reproduction

Male and female flowers are on separate plants, and the female flower is unique in that it grows on a filament that extends from the rhizome, and is surrounded by sessile leaves. The female flower produces 6 styles unlike many other *Halophila* species which have 3 styles. Little is known about the timing of flowering in this species.

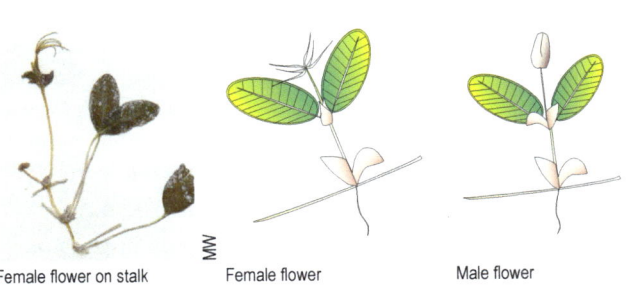

| Female flower on stalk | Female flower | Male flower |

Flowering Fruiting

| J | F | M | A | M | J | J | A | S | O | N | D |

Unknown in this region

Distribution

LC

Seed

Habitats

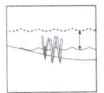 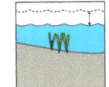

 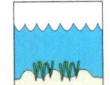

Key characteristics

- ▶ Leaves in pairs
- ▶ Leaf width 0.75–2 cm, length 2.5–7 cm
- ▶ Leaf cross-veins at 45°
- ▶ Endemic to southern Australia

Distinctive feature

- ▶ Female flower with 6 styles and on an erect stalk

Halophila decipiens

Ostenf.

Halophila decipiens has a global distribution. The very low degree of genetic divergence among plants across the globe suggests a relatively recent colonisation across oceans. It is an ephemeral and fast-growing species. Superficially, it appears like *H. ovalis*, and often occurs in patches adjacent to, or mixed with, *H. ovalis* and *H. spinulosa*. The species is more commonly found in subtropical and tropical environments, where it frequently occupies deepwater habitat (> 50 m), providing important forage for grazers, but also near reef and in sandy habitats. In temperate habitats (southern Australia and southern Mozambique in Africa), it is not observed in deepwater or reef habitats and is most commonly observed near the entrance of open estuaries.

Leaf pairs and rhizome of *Halophila decipiens*

Taxonomy and morphology

H. decipiens is clearly differentiated from other temperate species of *Halophila* by having serrated leaf margins and minute hairs on the surface of the leaves. The leaves are clearly borne on petioles (short stalks below the leaf blade), which arise directly from the rhizome. Leaves are up to 25 mm long and 6 mm wide.

Hydrocharitaceae

Reproduction

Both male and female flowers of *Halophila decipiens* are found on the same plant, produced at each reproductive node. The male flower develops first, producing ellipsoidal-shaped pollen with no exine. The female flower develops into small, sometimes green fruits, up to 6 mm in diameter, which may contain up to 30 small seeds (less than 0.5 mm diameter). This species sometimes behaves as an annual. A seed bank containing almost 177,000 seeds m^{-2} was recorded in the Hardy River estuary, Western Australia.

Male

Fruit

Seeds

Seed

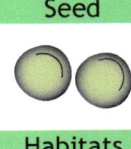

Habitats

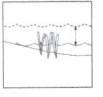
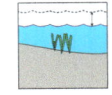
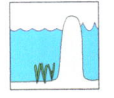
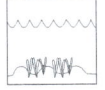

Flowering Fruiting

J F M A M J J A S O N D

Distribution

LC

Key characteristics

▶ Paddle-shaped leaves, the length longer than the width, with minute serrations along the edge
▶ Pairs of leaves on petioles that arise directly from rhizome
▶ Flowers of both sexes arise from the base of the same leaf pair, male flowers emerge first
▶ Widely distributed throughout temperate and tropical habitats

Distinctive feature

▶ Minute hairs on both sides of leaf blade

Halophila ovalis 'complex'

(R.Br.) Hook.f.

CR

Halophila ovalis has one of the widest environmental ranges of all the seagrasses, occurring in low salinity to hypersaline waters, from intertidal to deepwater (> 30 m) and from sheltered to highly dynamic conditions. It also exhibits significant morphological variation. It is frequently the first species to colonise disturbed areas, forming a single-species meadow but also forms patches in mixed-species meadows. In temperate regions, *Halophila ovalis* has little growth or even dies back in winter due to the low temperatures. It is commonly grazed by swans in estuarine systems. *Halophila ovalis* also forms a 'complex', which includes a number of species or sub-species for which the taxonomy is not clear at present.

KM

Leaf pairs of *H. ovalis*

Taxonomy and morphology

Halophila ovalis has pairs of leaves arising from the rhizome on a petiole. The leaves have > 4 cross-veins, an intramarginal vein and smooth edges. This species can be confused with *H. australis*, but its female flowers have 3 styles compared to 6 in *H. australis*. It can also be confused with *H. decipiens*, which has hairs on the leaf surface and serrated margins. The taxonomy of the *Halophila ovalis* 'complex' is discussed on the following pages.

Reproduction

In members of the *Halophila ovalis* 'complex', male and female flowers form on separate plants at reproductive nodes. The male flower produces 3 stamens, which are extended to the top of the leaves when the ellipsoidal pollen is released. The 3-pronged stigmas of the female flowers extend beyond the height of the leaves, maximising their opportunity to catch pollen in the water column. Flowering is often dense and seed set heavy, though the timing of flowering is more common in spring and summer in the southern temperate region, compared to year-round in the tropics. Seeds are small, yellow in colour with a hard outer ornamented seed coat.

Female

KV

Male

KV

Released pollen

KV

Seed

Habitats

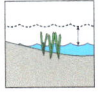 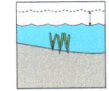
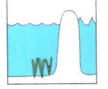 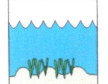
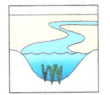

Flowering
Fruiting

J	F	M	A	M	J	J	A	S	O	N	D

Distribution

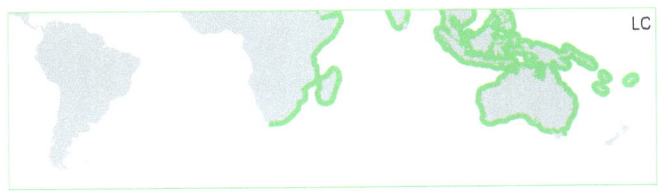

LC

Key characteristics

▶ Paddle-shaped leaves of variable proportions with smooth edges and no serration
▶ Pairs of leaves on petioles that arise directly from the rhizome
▶ Leaves sometimes have small red dots near the mid-vein
▶ Widely distributed throughout temperate and tropical regions

Distinctive feature

▶ Female flowers with 3 styles, attached to rhizome

Halophila ovalis 'complex'

To aid in understanding the taxa that have been described in the *Halophila ovalis* 'complex', we provide the key characteristics of those found in the southern temperate region. There are other members of this 'complex' that are found predominantly in tropical environments (see Waycott *et al*. 2004). In addition, recent taxonomic work, primarily on tropical species from the Asian region, has described new species in the *Halophila ovalis* 'complex' and molecular evidence shows that individuals with a similar genetic signature, indicating that they are the same taxa, are also found in temperate Australia. For this reason, we include *H. ovalis*, *H. ovata*, *H. minor* and *H. euphlebia* in the *Halophila ovalis* 'complex' present in the temperate Southern Hemisphere.

At present, *H. ovata* is not recognised as a legitimate taxonomic name by IPNI, although it is used in the seagrass literature and some taxonomists have suggested it should be named *H. gaudichaudii*. *Halophila euphlebia* is a recently proposed species within this 'complex' and some studies have identified samples with a similar genetic signature in Australia. Its distribution in the southern temperate region is not well known.

There are no clear morphological distinctions or diagnostic characters that separate these taxa, and the key characteristics of each described taxon often overlap in their ranges. Additional research is needed to resolve the phenotypic, ecological and genetic variation and confirm species boundaries in this species 'complex'.

Halophila ovalis (R.Br.) Hook.f.

Key characteristics
- ▶ Leaf cross-veins 10–12
- ▶ Leaf width 4–7 mm
- ▶ Space between veins and intramarginal vein 0.2–0.4 mm

Halophila euphlebia Makino

Key characteristics
- ▶ Leaf cross-veins 12–19
- ▶ Leaf width 5–15 mm
- ▶ Space between veins and intramargina vein 0.1–0.2 mm

Halophila ovata
(H. gaudichaudii) Gaucich.

Key characteristics
- ▶ Leaf cross-veins 4–8
- ▶ Leaf width 4–8 mm
- ▶ Space between veins and intramarginal vein 0.4–0.6 mm

Halophila minor (Zoll. Hartog

Key characteristics
- ▶ Leaf cross-veins 7–12, occasionally split veins
- ▶ Leaf width 3.5–6 mm
- ▶ Space between veins and intramarginal vein 0.15–0.19 mm

Halophila spinulosa

(R.Br.) Asch.

This species has a primarily tropical distribution but is found in warmer temperate waters of Western Australia. The unusual, fern-like form of *Halophila spinulosa* is distinctive among seagrasses but can lead to this species being confused with several species of green algae. In temperate regions, it is relatively scarce and, while there is no evidence that it occurs in deep habitat (> 20 m), it is frequently found in the deeper channels and basins of shallower habitat (< 10 m) and often associated with *Halophila ovalis*. Due to its complex growth form, *Halophila spinulosa* provides a structurally complex habitat for a variety of organisms where it is dominant. Dugongs are known to graze this species in Shark Bay.

Leaf pairs on vertical stem

Taxonomy and morphology

This species is distinctive and, once recognised, easy to remember. Leaves arise in opposite pairs from a vertical stem, forming a flattened, frond-like shape. The small, oblong-shaped leaves are minutely serrated along the edges, with an obvious central vein. The vertical stem, up to 30 cm long, may have over 20 leaf pairs, with old leaves dropping off near the base, leaving leaf scars. Young *Amphibolis* shoots may superficially appear like *H. spinulosa* but the leaves lack a flattened structure.

Hydrocharitaceae

Reproduction

Male and female flowers are on separate plants in *H. spinulosa*, both forming at the base of the leaves on the vertical stem. The pollen is ellipsoidal without an exine. When mature, the flask-shaped fruits are often left protruding from the vertical stem, containing up to 30 tiny seeds (around 0.5 mm diameter). The timing of flowering is not known in the southern temperate region.

Male

Female

Fruit

Seed

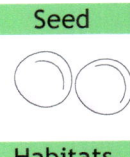

Habitats

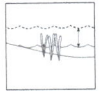 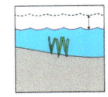

 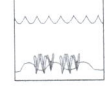

 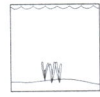

Flowering Fruiting

J	F	M	A	M	J	J	A	S	O	N	D

Unknown in this region

Distribution

Key characteristics

- All leaf pairs in a single plane, on vertical stem
- Small one-sided 'fold' at the base of each leaf
- Leaves have minute serrations on their edges
- Single flowers form at the base of leaves, may be many per vertical stem
- Central south Asian tropical distribution extending south to Jurien Bay, Western Australia

Distinctive feature

- Leaf pairs on opposite sides of vertical stem

Zosteraceae

The Zosteraceae is a seagrass family that contains one of the best known seagrass species, *Zostera marina* (eelgrass). This Northern Hemisphere, temperate species even grows in Alaska, seasonally covered by ice. Members of this family are primarily temperate, preferring cooler waters. However, two *Zostera* species in the southern temperate region are also found in tropical waters, *Zostera muelleri* in Australia and *Zostera capensis* in southern and eastern Africa. The taxonomy relating to the number of genera has been controversial. Here we recognise two genera in this family, *Zostera* and *Phyllospadix* (surf-grass). A previously well-recognised genus, *Heterozostera*, is included in *Zostera* following the work of Jacobs and Les (2009). Yet the recent *Flora of Australia* (Volume 39, Kuo 2011) does not reflect these changes and maintains *Heterozostera*. These generic boundaries remain unresolved but in this guide we adopt the treatment of Jacobs and Les.

Zostera species are commonly found in protected areas and estuaries. The extensive meadows they form are particularly important primary producers and provide a habitat for many other species of plants and animals. Contrasting with this are the habitats typically occupied by *Phyllospadix* species in the Northern Hemisphere. *Phyllospadix* is found on rocky shores with heavy wave action.

As a family the Zosteraceae are relatively easy to recognise if flowering structures are visible. The inflorescence is characteristic and contains both male and female flowers. The flowers are enclosed within a leaf structure, called a spathe, borne on a specialised shoot. The male flowers usually mature first to avoid pollinating their own female flowers. Once pollinated the female flowers form small seeds lined up within the spathe which, when mature, are released into the water column. These seeds are eaten by many animals common in seagrass meadows, including small prawns, and are consumed by dugong incidentally when grazing in *Zostera* meadows. Seeds are known to survive in sediments for up to one year.

Opposite top: *Zostera tasmanica*.
Opposite bottom: *Zostera muelleri*.

Zostera capensis

Zostera capensis occurs along eastern Africa, from Kenya to east of Cape of Good Hope, and in Madagascar and the Seychelles and is similar to *Z. muelleri*. In this temperate region, it is restricted to estuaries and lagoons and usually forms dense, monospecific meadows. It can be a component of multi-species meadows in subtropical and tropical areas. About 5 km² of *Zostera capensis* meadow have been recorded in 11 estuaries in South Africa. It is believed that many of these populations have suffered declines, due primarily to heavy flooding. In Mozambique, overharvesting of bivalves from meadows has had a devastating effect on seagrass cover and threatens both the meadows and the local human population's food resource.

Exposed at low tide in southern Mozambique

Taxonomy and morphology

The morphology of *Z. capensis* is highly variable. The species is distinguished from others in temperate regions by having rounded leaf tips and a rhizome less than 3 mm in diameter. In subtropical habitats, the leaf form may be similar to *Cymodocea* spp. but differs in the leaf sheath. The closely related species, *Zostera muelleri*, has a similar morphology but is found in Australia and New Zealand.

Zosteraceae

Reproduction

In *Zostera capensis* male and female flowers are within a single inflorescence that is enclosed in a spathe (specialised leaf), containing up to five pairs of male and female flowers. The shoot on which flowers form may have up to seven spathes on it. Seeds are a shiny, red-brown nut, 2–2.5 mm long, that can form a seed bank. Not much is known about the flowering biology of *Z. capensis*.

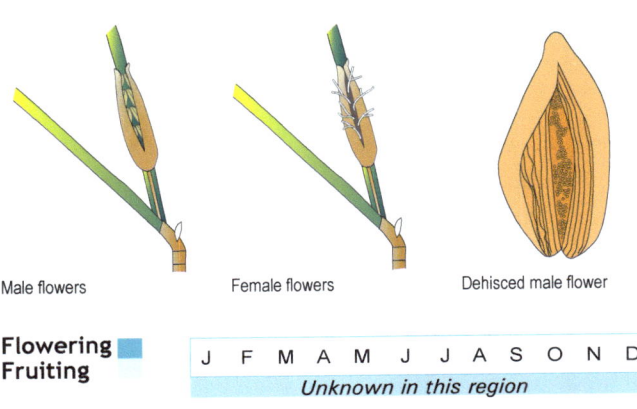

Male flowers   Female flowers   Dehisced male flower

Seed

Habitats

 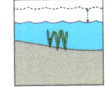
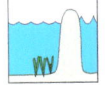 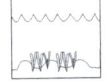

Flowering Fruiting

J	F	M	A	M	J	J	A	S	O	N	D

Unknown in this region

Distribution

VU

Key characteristics

▶ Leaves arise directly from rhizome, leaving a persistent membranous sheath
▶ Rhizome usually dark brown or yellow
▶ Cross-veins visible in leaf
▶ Flowers, both male and female, borne in a spathe that appears as a thickened leaf sheath

Distinctive feature

▶ Rhizome < 3 mm diameter, Africa

Zostera muelleri

Irmisch ex Asch.

JG

Until recently, *Zostera muelleri* was known from temperate, estuarine systems of Australia but now also includes *Z. novazelandica* from New Zealand and *Z. mucronata* from Australia. Molecular and morphological studies confirmed they are the same widespread and morphologically variable species. Typically marine, it can tolerate some freshwater inputs and occurs in estuaries, shallow bays and the intertidal. It often forms mono-specific beds but can grow with *Ruppia*, *Halophila* and *Lepilaena*. *Zostera muelleri* meadows support a diverse and productive community, providing food for wading birds and nursery areas for commercial fish and shrimp species. Significant losses have been recorded in Port Phillip Bay and New Zealand due to sedimentation, turbidity and direct disturbance of habitat.

KM

Estuarine plants

Taxonomy and morphology

Zostera muelleri is synonymous with *Z. mucronata*, *Z. capricorni* and *Z. nova-zelandica*. Its morphology is highly variable and it is distinguished from other temperate seagrasses by its rounded leaf tips and thin rhizomes (< 3 mm diameter). Young rhizomes can be yellow, while leaves exposed to high light can be reddish. This species can be confused with young plants of the *Z. tasmanica* 'complex' and with *Z. capensis*, which is only found in Africa.

Reproduction

Like *Z. capensis*, flowers are formed in an inflorescence that is enclosed in a spathe. Up to 6 spathes can occur on each shoot, each containing 4–12 pairs of male and female flowers. The intensity of flowering is positively related to the size of the plant. The male flowers mature before the female flowers and peak flowering times vary with location. As the flowering shoot matures, it darkens and eventually breaks off. The shoot may float away with the seeds still enclosed, later dropping the small, hard-coated nuts to the sediment.

Flowering spathe

Female flowers

Male flowers

Seed

Habitats

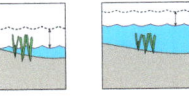
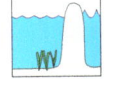
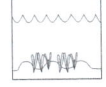

**Flowering
Fruiting**

J F M A M J J A S O N D

Distribution

LC

Key characteristics

▶ Leaves arise directly from rhizome, leaving a persistent membranous sheath
▶ Rhizome usually dark brown or yellow
▶ Cross-veins visible in leaf
▶ Flowers, both male and female, borne in a spathe that appears as a thickened leaf sheath

Distinctive feature

▶ Rhizome < 3 mm diameter, Australia and New Zealand

Zostera tasmanica 'complex'
(formerly *Heterozostera tasmanica*)

FVR

Zostera tasmanica (formerly *Heterozostera tasmanica*) has recently been split into four species by Kuo in 2005. The group of species that were formerly *Z. tasmanica* occurs in a wide range of habitats, including estuaries, coastal lagoons and embayments of varying degrees of exposure, but are typically more marine than *Zostera muelleri*. In more sheltered environments it can form mono-specific meadows but also forms mixed meadows, commonly with *Halophila ovalis*. In Australia, its ecological significance is difficult to quantify as it is commonly a sub-dominant in multi-species meadows, but in Chile it forms large mono-specific meadows that are important fishery habitat, including that of the Chilean scallop fishery.

FVR

Port Phillip Bay

Taxonomy and morphology

We present the broader species concept of *Z. tasmanica* 'complex' here, the newer species are discussed later. Unlike other *Zostera* species, *Z. tasmanica* forms erect shoots that may become wiry with age. The vertical stem retains fragments of old leaves, differentiating it from *Amphibolis* and *Thalassodendron*, as does its narrower leaves. This species may be confused with *Zostera muelleri* when young.

Zosteraceae

Reproduction

Zostera tasmanica also has male and female flowers on the same plant, enclosed in a specialised spathe with up to 14 flowers per spathe. Flowers develop and mature in summer and spring, although there can be variations between years in the intensity of flowering. With the high density of flowers, prolific seed production is possible, though only an average of 70 seeds m^{-2} has been recorded. The specialised reproductive shoots can break off and disperse the seeds. Seeds are dark-brown to shiny black in colour with a hard outer coat.

Spathes in canopy

Thickened spathes

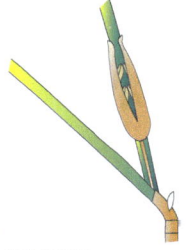

Male flowers

Seed

Habitats

 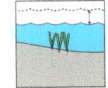

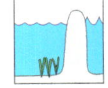 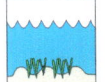

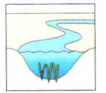

Flowering
Fruiting

| J | F | M | A | M | J | J | A | S | O | N | D |

Distribution

LC

Key characteristics

- Leaves borne on end of erect stem that is rarely branching
- Flowers, both male and female, borne in a spathe
- Leaf width < 3 mm, length > 8 cm
- Leaf cross-veins conspicuous
- Found throughout temperate Australia and Chile

Distinctive feature

- Old leaf sheaths present on vertical stem

Zostera tasmanica 'complex'

Significant morphological variation has been observed across the range of *Zostera tasmanica*. This variation has recently been categorised into three new species plus *Z. tasmanica*. One species is found only in two locations in Chile, South America. The remaining three species are differentiated on the basis of wiry erect stems (*Z. nigricaulis*) and the number of flowers in each reproductive structure (spathe): those with few female and male flowers are *Z. tasmanica*, those with many flowers, *Z. polychlamys*. These characters are subtle and not always possible to observe in field-collected specimens. We provide diagnostic features below, but it should be noted that when young or lacking an erect stem this species will be difficult to discriminate from *Zostera muelleri*. *Zostera tasmanica* species will typically be found in oceanic rather than estuarine habitats. In contrast, *Z. muelleri* is typically found in estuarine habitats.

Zostera tasmanica   G.Martens ex Asch.

Key characteristics
- ▶ Wiry, erect stems, not black, leaf tip rounded with minute slit, reproductive structures with up to 4 female and male flowers
- ▶ Found in Victoria and Tasmania, Australia

Zostera nigricaulis   (J.Kuo) S.W.L.Jacobs & Les

Key characteristics
- ▶ Wiry, erect stems, black, branching at top, notch in leaf tip
- ▶ Found in temperate Australia

Zostera polychlamys   (J.Kuo) S.W.L.Jacobs & Les

Key characteristics
- ▶ Wiry, erect stems, not black, leaf tip squared off with shallow notch, reproductive structures with up to 12 female and male flowers
- ▶ Found in WA and SA, Australia

Zostera chilensis   (J.Kuo) S.W.L.Jacobs & Les

Key characteristics
- ▶ Wiry, erect stems, black, rarely branching at top, notch in leaf tip
- ▶ Found in Chile

Potamogetonaceae

There are few members of the marine Potamogetonaceae, previously in their own family the Zannichelliaceae. There were three genera in the Zannichelliaceae, *Lepilaena* (Australia and New Zealand), *Zannichellia* (cosmopolitan) and *Althenia* (Mediterranean). Members of this group occur in coastal habitats ranging from wetlands, coastal marshes and lagoons to estuaries and sheltered embayments. Most species of this global group are known to tolerate a wide range of salinities from freshwater to hypersaline.

Plant morphology among the southern temperate members (i.e. *Lepilaena*) of this family is typically delicate, shoots are narrowly linear and branching, the flowers and seeds being formed near the shoot tips. In the field, members of this family are most likely mistaken for other seagrasses or aquatic plants, typically *Ruppia* or *Potamogeton*. For example, without reproductive structures, *Lepilaena* is very difficult to discriminate in mixed stands of *Ruppia*, which appear to be common in some protected embayment and estuarine habitats. At least two species of *Lepilaena* co-occur with other seagrasses in temperate Australia.

The marine Potamogetonaceae represent the fourth lineage in the Alismatales that have become adapted to the marine environment. Most species appear to be highly ephemeral and are recorded as producing prolific numbers of seeds, relying on these to recover from stressful environmental conditions. However, for most species of *Lepilaena*, we have a poor understanding of their distribution and ecology.

Lepilaena growing in an estuary

Lepilaena

J.Drumm. ex Harv.

Members of this genus are not usually considered as 'seagrasses', however, several species of *Lepilaena* co-occur with other, well-known seagrasses such as *Zostera muelleri* and *Halophila ovalis*. Most species of *Lepilaena* appear to tolerate marine salinity, although as they are typically recorded from more brackish to fresh conditions, we might view this group as a transitional marine to freshwater group. As they occupy habitats with variable conditions, *Lepilaena* species are often ephemeral and produce prolific numbers of seeds facilitating recovery. The general ecology of these plants is poorly known.

Fine leaves with clusters of flowers

Taxonomy and morphology

This genus is only found in Australia and New Zealand. Only three of the five species have been recorded at marine salinities— *Lepilaena preisii*, *L. cylindrocarpa*, *L. marina*. Plants are difficult to distinguish from *Ruppia* spp. if no reproductive structures are observed. However, *Lepilaena* has a ligule, or membrane between the leaf sheath and the new leaf blade which *Ruppia* does not. Leaves are flattened (typically) to circular in cross-section; leaf tips have a distinctive point. The overall plant form is variable although more delicate than most species it may co-occur with.

Potamogetonaceae

Lepilaena

Reproduction

The species of *Lepilaena* can have male and female on the same or different plants. Flowers can be solitary or aggregated together in an inflorescence and form at the end of a shoot appearing as clusters. Pollination occurs underwater. Fruits are green when immature to dark brown when older, having a distinct beak on the seed. Female flowers have an open funnel shaped stigma which darkens after pollination.

Female flowers

KM

Developing fruit

KM

Flowers near top of canopy

KM

Seed

Habitats

 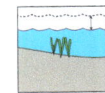

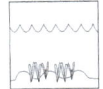

Flowering
Fruiting

J	F	M	A	M	J	J	A	S	O	N	D

Unknown in this region

Distribution

DC

Key characteristics

- Fine, cylindrical leaves
- Pointed leaf tips
- Leaves < 1 mm wide
- Larger plants often much branched
- Widespread through temperate and tropical regions

Distinctive feature

- Flowers in clusters at end of shoot

Southern Temperate Seagrasses ❧ 99

Bibliography

General

Larkum A *et al.* (2006). *Seagrasses: Biology and ecology and conservation.* Springer, Dordrecht.

den Hartog C (1970). *The sea-grasses of the world.* North-Holland Publishing, Amsterdam.

Green EP, Short F (2003). *World atlas of seagrasses.* University California Press, Berkeley.

Waycott M *et al.* (2004). *A guide to the tropical seagrasses of the Indo-West Pacific.* James Cook University, Townsville.

Butler A, Jernakoff P (1999). *Seagrass in Australia: Strategic review and development of a research and development plan.* CSIRO Publishing, Melbourne.

Habitats and Bioregions

Carruthers TJB *et al.* (2007). Seagrasses of south-west Australia: A conceptual synthesis of the world's most diverse and extensive seagrass meadows. *Journal of Experimental Marine Biology and Ecology* **350**: 21–45.

Short FT *et al.* (2007). Global seagrass distribution and diversity: A bioregional model. *Journal of of Experimental Marine Biology and Ecology* **350**: 3–20.

Currents

Pearce AF (1991). Eastern boundary currents of the southern hemisphere. *Journal of Royal Society of Western Australia* **74**: 35–46.

Nybakken J, Bertness M (2005). *Marine biology: An ecological approach.* Benjamin Cummings, San Francisco.

Ecology

Heminga MA, Duarte CM (2000). *Seagrass ecology.* Cambridge University Press, Cambridge.

Valentine JF, Duffy JE (2006). The central role of grazing in seagrass ecology. In: Larkum *et al.* pp. 461–501.

Kendrick G *et al.* (2012). The central role of dispersal in the maintenance and persistence of seagrass populations. *BioScience* **62**: 56–65.

Hughes AR *et al.* (2009). Associations of concern: Declining seagrasses and threatened dependent species. *Frontiers in Ecology and the Environment* **7**: 242–246.

Connell S, Gillanders B (2007). *Marine ecology.* Oxford University Press, Melbourne.

Evolution and Taxonomy

Les DH *et al.* (1997). Phylogenetic studies in Alismatidae, II - evolution of marine angiosperms (seagrasses) and hydrophily. *Systematic Botany* **22**: 443–463.

den Hartog C, Kuo, J (2006). Taxonomy and biogeography of seagrasses In: Larkum *et al.* pp. 1–23.

Waycott M *et al.* (2006). Seagrass evolution, ecology and conservation: A genetic perspective. In: Larkum *et al.* pp. 25–50.

Cronquist A (1981). *An integrated system of classification of flowering plants.* Columbia University Press, New York.

Womersley HBS (1984). *The marine benthic flora of Southern Australia.* State Government, Adelaide.

Wilson A (ed.) (2011) *Flora of Australia Vol 39. Alismatales to Arales.* Flora of Australia Series. Australian Biological Resources Study (ABRS)/CSIRO Publishing, Melbourne.

Arber A (1920). *Water plants: A study of aquatic angiosperms.* Cambridge University Press, London.

Jacobs SWL, Les DH (2009). New combinations in *Zostera* (Zosteraceae). *Telopea* **12**: 419–423.

Kuo J (2005). A revision of the genus *Heterozostera* (Zosteraceae). *Aquatic Botany* **81**: 97–140.

Kuo J (2011). Zosteraceae. In: *Flora of Australia Vol. 39. Alismatales to Arales.* Australian Biological Resources Study/CSIRO Publishing, Melbourne.

Growth and Reproduction

Ackerman JD (2006). Sexual reproduction of seagrasses: Pollination in the marine context. In: Larkum *et al.* pp. 89–109.

Duarte CM *et al.* (2006). Dynamics of seagrass stability and change. In: Larkum *et al.* pp. 271–294.

Marba N, Walker DI (1999). Growth, flowering, and population dynamics of temperate Western Australian seagrasses. *Marine Ecology Progress Series* **184**:105–118.

Human interactions with seagrass

Orth RJ *et al.* (2006). A global crisis for seagrass ecosystems. *Bioscience* **56**: 987–996

Duarte CM *et al.* (2008). Seagrass ecosystems: their global status and prospects. In: Polunin NVC (ed.) *Aquatic ecosystems.* Cambridge University Press.

Costanza R *et al.* (1997). The value of the world's ecosystem services and natural capital. *Nature* **387**: 253–260.

Short FT (2011). Extinction risk assessment of the world's seagrass species. *Biological conservation* **144**: 1961–1971.

Nellemann C *et al.* (2010). *Blue carbon. The role of healthy oceans in binding carbon.* United Nations Environment Programme, GRID-Arendal, Norway.

Pictorial glossary

There are several different morphological forms of seagrass. To help with identifying seagrass species, some of the forms and features that occur are illustrated below.

Leaves strap-like

Arise directly from rhizome

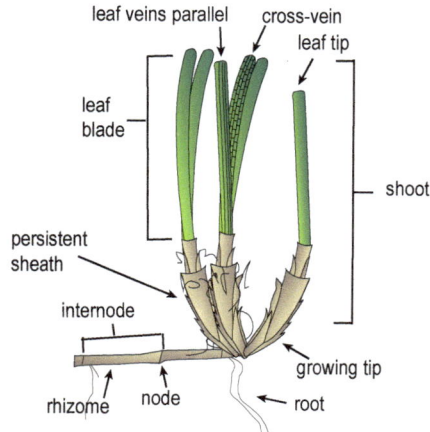

Arise from vertical stem

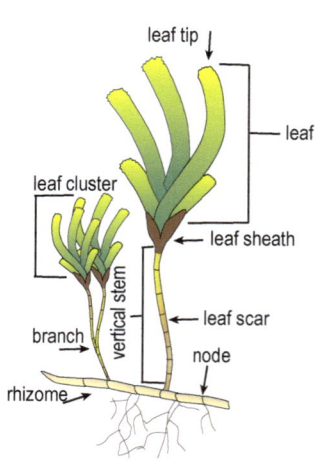

Leaves oval to oblong

In pairs attached to the rhizome

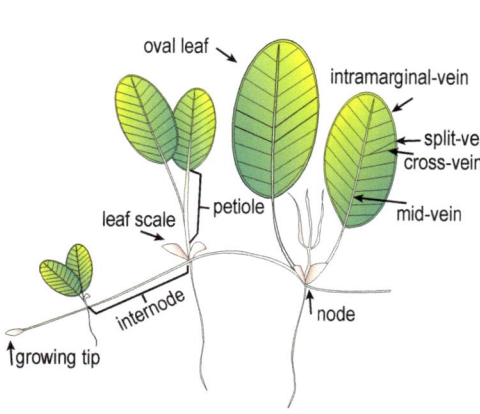

Attached to vertical stem

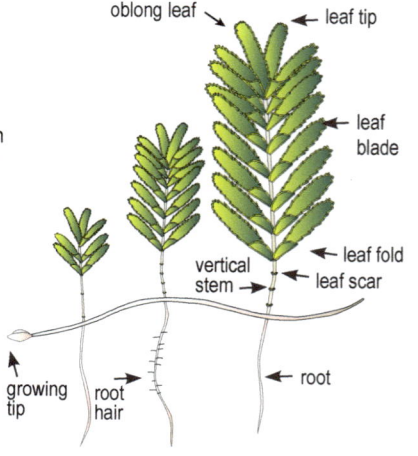

Flowering structures also vary among seagrasses. Some of the key flowering and seedling features are illustrated below.

Leaves cylindrical

Reproductive structures

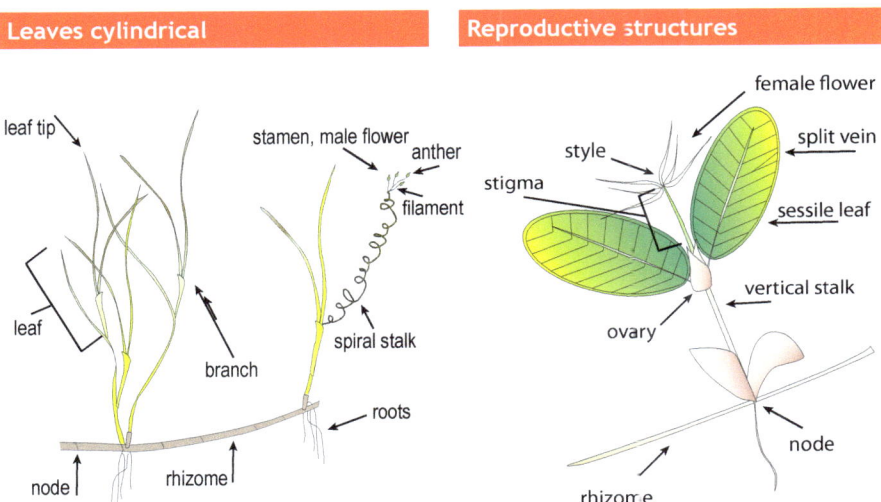

Reproductive structures

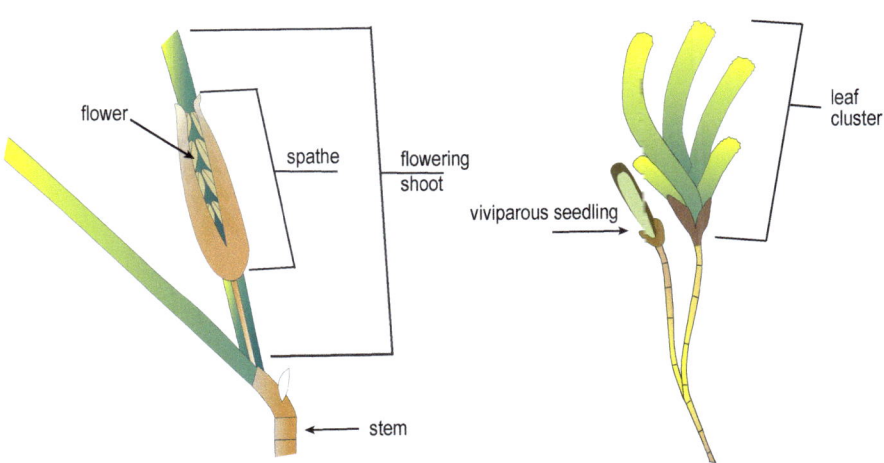

Glossary

angiosperms	A group of flowering plants distinguished by true flowers
annual	A plant that germinates, grows, flowers, sets seeds and dies within a year
anther	Pollen-bearing (male) part of a flower; usually the terminal portion of the stamen (p103)
bi-dentate	Having 2 teeth-like projections
bioregion	A geographic region that is defined by the types of organisms that live there
bisexual	Male and female parts in one individual
bract	Leaf-like structure at the base of an inflorescence or flower
branch	Where a leaf, shoot, stem or rhizome splits into two (p102)
bristle	A stiff structure, such as a hair or projection
bullose	Distended surface with a blistered or puckered appearance
carpel	Central part of the flower enclosing the ovules (female)
clonal growth	Multiple organisms produced from one parent by vegetative means; individual plants thus formed may later grow independently
comb	Referring to the hard, comb-like structure at the base of an *Amphibolis* seedling that aids in attachment of the seedling to the ground
complex	Referring to a related group of species where the boundaries between the species are unclear, indicating there is uncertainty in the taxonomy
convergent	Appearing to have a similar form but different ancestry
cryptic	Not visually obvious or, in reference to species, species that can not be visually distinguished
cyme	An inflorescence in which each flower, in turn, is formed and then opens at the tip of a growing axis
dentate	Having toothed-like projections
dioecious	Having male and female flowers on different plants
drupe	Fleshy fruit derived from a single carpel, usually containing a single seed
endemic	Unique to a given region
epiphyte	Organisms that grow on seagrass, e.g. algae
ephemeral	A plant with a short life cycle; generally does not persist in a habitat for an extended period
estuary	Body of water, which links river and ocean, generally with fluctuating salinity; adj. estuarine
evolutionary lineage	Shared ancestry of a particular group of plants
exine	Outer layer of the wall of a pollen grain
family	Taxonomic group between order and genus, family names end in -*aceae*
fibre (n)	A fine, slender thread of persistent tissue found in plants; adj. fibrous
flowering plant	A plant where ovules, and then seeds, develop within an enclosed ovary (angiosperm)
fruit	The developed ovary of a flowering plant
genus pl. **genera**	Main taxonomic group above species and below family
germination	The beginning of the growth of a plant
green algae	Freshwater or marine organism, either unicellular or multicellular; seaweed (Chlorophyta)
habitat	The environment of an organism, defined by physical, chemical and biological features

hairy	Consisting of or resembling hair
hermaphroditic	Having male and female parts within the same flower
hypersaline	Salinity greater than seawater
inflorescence	The group or arrangement of flowers on a plant
internode	The part of a plant stem between successive nodes (p102)
intertidal	Region that is successively exposed to air and inundated with seawater with tidal movement
lateral teeth	Teeth or dentations that are directed sideways, in seagrasses usually on leaves
leaf	The photosynthetic organ of a plant, of various shapes and features

overall shape:
cylindrical (p103)   oval (p102)   oblong (p102)   strap-like (p102)

shape in cross-section:
round ◯   concave ⟋   convex ⟋   biconvex ⟋   flattened ⟋

leaf blade	The main part of the leaf, distinct from the sheath and petiole (p102)
leaf cluster	A group of leaves arranged together, in seagrasses usually on a vertical stem (p102)
leaf fold	Base of leaf; a structure found only in *Halophila spinulosa* (p102)
leaf hair	A hair that grows from the leaf surface; a single cell or single row of cells
leaf margin	The edge of a leaf

serrated   smooth ⌒

leaf pair	Two leaves arising from the same location on a stem or rhizome; typical of most *Halophila* species (p102)
leaf scale	A thin membranous flap of tissue that generally covers the leaves in early leaf development, found commonly in *Halophila* (p102)
leaf scar	A scar on a stem or rhizome marking where a leaf was once attached

open   closed ▰

leaf sheath	The lower part of a leaf which clasps the stem, or a wing-like extension to the margins of the petiole which wrap around the stem; sheaths may persist after the leaf dies (p102)
leaf tip	End of leaf with a variety of shapes, for identification use young leaves, as old leaves are easily damaged

flat ⎚   concave ⊓   pointed ⟋   serrated ⫲
rounded ⌒   oblique ⊐   notched ⌄   truncate ⊓

leaf vein	Strands of vascular tissue in leaves that transport water, nutrients and photosynthetic products; the orientation of veins is used for identification

cross-vein - perpendicular to leaf length, i.e. across the leaf (p102)
intramarginal-vein - around inside edge of leaf (p102)
mid-vein - prominent central vein (p102)
parallel-vein - two or more veins parallel and along the length of the leaf (p102)
split-vein - cross-veins that fork (p102)

leathery	Like leather in appearance, tough and flexible
ligule	A membranous structure that may occur between a leaf and its sheath, or a strap-like appendage from other plant parts
linear	Narrow, compared to length, with parallel edges
lineage	A group of species connected by a common ancestor
marginal serrations	Of leaves, finely toothed edges
marine plant	Any free living organism capable of photosynthesis submerged in the ocean
membranous	A flattened and thin structure, usually smooth
meristem	The tissue from which new cells are produced, often at the base of leaves in seagrasses
monocotyledon	A flowering plant that has a single seed leaf in the embryo; this group does not form woody stems and includes grasses and orchids. Also monocot
monoecious	Having separate male and female flowers but on the same individual plant
monospecific	Formed from a single species
morphological plasticity	Where a species has a wide variation in the form and size of plant parts, not due to genetic differences, and often stimulated by environmental factors
node	Part of a plant stem or rhizome where leaves and buds arise, and where branching occurs (p102)
nut	A hard, dry, one-seeded fruit, which doesn't open at maturity; generally produced from a compound ovary
opposite	A leaf arrangement where two leaves or buds arise at the same position but on opposite sides of the stem
ovary	Of the female part of a flower, the basal portion of a carpel that surrounds and protects the ovule(s) (p103)
ovule	The egg cell of a female flower
pedicel	The stalk of an individual flower
perennial	A plant that grows for longer than two growing seasons; cf. annual
persistent leaf sheath	Leaf sheath that remains attached to the plant after the leaf has died; can be membranous or with soft fibres (hairy)
petiole	Stalk of a leaf (p102)
phenotype	The appearance of an organism resulting from its genetic make-up and interaction with the environment
photosynthesis	Process where plants use solar energy, water and carbon dioxide to create carbohydrates, which enables growth
phylogeny	The evolutionary history of a group of organisms estimated through analysis of traits
phytoplankton	Photosynthetic algae that live freely, suspended in the water
pollen	Male reproductive cells of seed plants (shed from anthers)
pollination	The transfer of pollen from the anther to the stigma; may be between flowers of the same or different plants
recruit	An individual plant that starts to grow in a new area, from seed or vegetative fragment
recurved	Curved backward or downward

rhizome	An underground stem, usually growing horizontally; they can be fragile, thick and starchy or almost woody; has scars where leaves were attached (p102)
rhizome mat	An underground, interwoven network of rhizomes
root	A plant part that grows down into the sediment from the node of the rhizome, anchors the plant and absorbs water and nutrients (p102)
root hair	Single celled, a fine projection from a root (p102)
seed	Part of a plant, product of a fertilised ovule, containing an embryo, a hard seed coat and food reserves, can persist for extended periods of time
seed bank	An accumulation of dormant seeds in the sediment that may germinate at a later time
serrated	Toothed or saw-like
sessile	Attached directly without a stalk
species	The basic taxonomic unit
shoot	Unit of foliage containing a number of leaf blades and a leaf sheath (p102)
spathe	A large bract at the base of an inflorescence that completely encloses it (p103)
spike	An upright stem bearing flowers without stalks
stable meadows	Meadows that persist for years to decades; cf. ephemeral meadows
stalk	A slender, supporting or connecting part of a plant (p103)
stamen	The pollen-bearing organ of a flower, consisting of a filament and anther (p103)
stem	Vertical plant part in some species with leaves attached; develops from the rhizome; has scars where leaves were attached (p102)
stigma	The terminal part of the style in a female flower that receives pollen (p103)
style	In a female flower, the slender column of tissue arising from the top of the ovary and through which the pollen tube grows, ending in a style
substrate	The base or material on which a plant grows
subtidal	Region that is always inundated with seawater, never exposed to air, irrespective of tidal movement
taxon pl. taxa	A group or category, at any hierarchical level, for classifying organisms, such as family, genus, species, subspecies
taxonomy	The science of describing, classifying and naming organisms
tri-dentate	Having three teeth-like projections
upwelling	The movement of less-dense surface water offshore and its replacement by cold, dense water from below the surface
vascular tissue	Plant tissue that consists of vessels and conducts materials e.g. water, photosynthates
vegetative	All plant parts not directly involved with sexual reproduction
viviparous	Of seeds, germinating while still attached to the parent plant (vivipary) (p103)

Contributors

Authors

Michelle Waycott, School of Earth and Environmental Sciences, The University of Adelaide, Australia, michelle.waycott@seagrassonline.org

Kathryn McMahon, Centre for Marine Ecosystems Research, Edith Cowan University, Australia, k.mcmahon@ecu.edu.au

Paul Lavery, Centre for Marine Ecosystems Research, Edith Cowan University, Australia, p.lavery@ecu.edu.au

Paul, Michelle and Kathryn

Photographers

AC:	Ainsley Calladine	JH:	John Huisman
AG:	Adam Gartner	JS:	John Statton
BM:	Britta Munkes	JV:	Jennifer Verduin
CC:	Catherine Collier	KM:	Kathryn McMahon
CR:	Chris Roelfsema	KV:	Jent Kornelis van Dijk
DK:	Diana Kleine	MF:	Meredith Ferdie
FS:	Fred Short	MT:	Martin Thiel
FVR:	Fisheries Victoria Research	MW:	Michelle Waycott
GI:	Graeme Inglis	PL:	Paul Lavery
GK:	Gary Kendrick	RC:	Robert Coles
JC:	Joel Creed	RH:	Renae Hovey
JG:	Justin Gilligan	WLL:	Warren Lee Long

Artwork

Cover artwork: *Seagrass meadow*
Medium: Oil on canvas
Australian artist: Tony Davis, Bridgetown, WA, 2009
Seagrass botanical illustrations: *Temperate seagrasses*
Medium: Watercolours
Australian artist: Elizabeth Rippey, Perth, WA, 2009

Acknowledgements

Ainsley Calladine for assistance with conceptual diagrams
Diana Kleine and **Tirza Abb** for graphic design
Family, friends and colleagues for reviewing assistance, particularly **Juergen Kellermann** and **Mike van Keulen** for their full scientific review.

Facing page: *Amphibolis antarctica.*

www.ingramcontent.com/pod-product-compliance
Lightning Source LLC
Chambersburg PA
CBHW040747010626
45792CB00028B/2067